Springer Geography

The Springer Geography series seeks to publish a broad portfolio of scientific books, aiming at researchers, students, and everyone interested in geographical research.

The series includes peer-reviewed monographs, edited volumes, textbooks, and conference proceedings. It covers the major topics in geography and geographical sciences including, but not limited to; Economic Geography, Landscape and Urban Planning, Urban Geography, Physical Geography and Environmental Geography.

Springer Geography—now indexed in Scopus

More information about this series at https://link.springer.com/bookseries/10180

Ilaria Tombolini · Jesús Rodrigo-Comino ·
Luca Salvati

Land Quality and Sustainable Urban Forms

Changing Landscapes and Socioeconomic Structures of European Cities

Springer

Ilaria Tombolini
Department of Architecture and Planning
Sapienza University of Rome
Rome, Italy

Luca Salvati
Faculty of Economics
Department of Methods and Models for Economics
Territory and Finance (MEMOTEF)
Sapienza University of Rome
Rome, Italy

Jesús Rodrigo-Comino
Facultad de Filosofía y Letras
Departamento de Análisis Geográfico Regional y Geografía Física
University of Granada
Granada, Spain

ISSN 2194-315X ISSN 2194-3168 (electronic)
Springer Geography
ISBN 978-3-030-94731-6 ISBN 978-3-030-94732-3 (eBook)
https://doi.org/10.1007/978-3-030-94732-3

This Springer imprint is published by the registered company Springer Nature Switzerland AG
The registered company address is: Gewerbestrasse 11, 6330 Cham, Switzerland

Preface

In the panorama of studies related to the ability of lands to support both natural processes and agricultural production activities, this book introduces a still unexplored or under-studied theme: the relationship between urban sprawl in its various forms and land quality.

The first part of the book is dedicated to the motivations and the theoretical premises from which the research originates, connected to the concept of land and those of sustainable urban form. The theme of recent urbanization and the settlement dynamics that ensue and that pose a series of questions on the transformations that the environment and the landscape are facing is deepened. The second part concerns the complex path toward a sustainable use of land, both in terms of institutional and regulatory measures, and in terms of knowledge and understanding of soil degradation processes. Taking into consideration the experiences and researches dealt with on land degradation, soil quality, interpretative methods and analysis approaches toward these issues, we want to make explicit the link between the meanings of the terms land, soil, landscape, land degradation, soil consumption, land quality and soil sealing. The English term "land", used in most compound words, allows to refer not to a specific resource (for example the soil), but to a multifunctional system of resources, which interacts with other ecosystems, generating flows of energy and matter. In addition to the terms and phenomenologies, knowing also the policies that make the attempt to achieve a sustainable use of land and that focus on the urban context can be a good starting point in understanding the orientation that contemporary society and culture have toward the conception of the soil as a resource. At the same time, identifying any gaps that exist between the legislative instruments for soil protection and the aspects related to land quality, which are not considered by them, is the engine for developing new investigations on how to deal with land degradation issues.

This book focuses on the Mediterranean area and, in the third chapter, it is better clarified because in this part of Europe relationships between settlement dynamics and land quality are fragile ecosystems and diffused both from a biological and socio-economic point of views. We find landscapes that are particularly sensitive to land degradation processes (subject to land degradation, considered the antipodes of land quality) and which in recent decades have been particularly affected by anthropic pressure. Recent urban transformations have in fact concerned not only the urban nuclei but also and above all the vast areas that surround them, accentuating the processes of degradation and determining the transformation of traditionally compact cities toward a more dispersed form. Some settlement, morphological and socio-economic characteristics resulting from urban sprawl are studied in depth for the case of three metropolises deemed emblematic of this transition (Rome, Barcelona and Athens), through information deducible from existing literature.

In the fourth part, an unpublished analysis is presented concerning 76 metropolitan areas representative of southern Europe. The methodology used in this analysis is based on the relationship that exists between soil sealing (or soil waterproofing) and land degradation (or land degradation) aimed at an interpretation, at the metropolitan scale, of how in southern Europe the pattern of Urbanization (compact, dispersive, intermediate) affects the land's ability to support both natural processes and agricultural production activities in a diversified way. In particular, the data on land quality and data on land use were considered together in order to analyze the processes of urban growth and the occupation of productive land for a very large area that includes Greece, Italy, Spain, Portugal and some Balcanic countries.

Despite some initiatives already undertaken in favor of the soil, both from a planning point of view and in terms of good practices, it is noted that the lack of a common strategic framework with clear objectives and shared parameters at European Union level has not yet allowed to fully exploit the experiences gained and the potential of the political guidelines of the States in which the metropolises covered by this study fall. There is still a long way to go, in terms of sharing, integration and definition of strategies aimed at achieving certain targets. A necessary and innovative look toward land quality could help to consider the protection of the soil as a whole, even at the planning level.

Some key words are highlighted in the text, the meaning of which is taken up and deepened in a glossary, placed immediately after the conclusions. These words are grouped into seven thematic categories, which constitute the guiding thread along which the text develops: "resources or resource systems", "ownership/state of resources", "processes/phenomena", "programs and products for interpretation of properties and processes", "initiatives in favor of land", "space units", "indices and statistics".

Rome, Italy — Ilaria Tombolini
Granada, Spain — Jesús Rodrigo-Comino
Rome, Italy — Luca Salvati

Contents

About the Authors

Dr. Ilaria Tombolini is a Ph.D. who currently works in a cartography bureau. She is a biologist M.Sc. from the University of Roma Tre and Ph.D. in Architecture and Planning from La Sapienza University. She wrote more than 10 articles on ISI impact factor journals.

Dr. Jesús Rodrigo-Comino held a Ph.D. at the University of Málaga (Spain) in Geography in 2018. Currently, he is a member of the Department of Regional and Physical Geography at the University of Granada (Spain). He has written and edited books related to land management and environmental processes, presented several oral conferences and posters in international meetings, and published more than 170 indexed peer-reviewed papers (Scopus/WOS). He is editor-in-chief of Air, Soil and Water Research (SAGE). Also, he works a reviewer for more than 120 international indexed journals. His major topics are land degradation, soil geography and regional geography.

Dr. Luca Salvati, M.S., Ph.D. is an expert in Regional Statistics, Demography, Economic Geography and Urban Planning. He is an aggregate Professor of Economic Statistics at the University of Macerata—Department of Economics and Law, dealing mainly with institutional and economic statistics, statistics for the territory, theory of indicators, sustainable development and urban growth. He held a Ph.D. in Economic Geography and was a former staff researcher of the National Statistics Institute (ISTAT) and the Council for Agricultural Research and Economics (CREA) in Italy. He is editor-in-chief of Geography (mpdi). He has published more than 400 articles in English, 30 books, and many thematic essays, also with specific contributions on multivariate statistics and spatial analysis.

Chapter 1
Introduction

Abstract In the face of an increasingly growing population globally and, consequently, a demand for quality food in continuous growth, the maintenance of soil fertility and their ability to provide goods and services is a topic of particular interest in the international scientific debate. In addition to their production function, soil resources contribute to those often neglected regulatory and support services, on which not only food requirements but also other sources of psycho-physical well-being depend; the protection and enhancement of soil resources are also generators of the cultural and aesthetic values of the landscape associated with ecosystems in a good environmental state. An issue related to this theme, probably still unexplored or not studied in depth in the scientific literature, is represented by the relationship between urban sprawl and land quality, i.e. the ability of the soil to support agricultural productivity and/or natural vegetation. Through this question and the resulting answer, we want to propose a reflection on the need for a change or a passage, which should probably also be reflected in planning practices, from a vision that works only on sealing toward a conception that considers the soil as a living entity, capable of hosting and supporting a varied multiplicity of organisms. We, therefore, propose an expansion of interpretative categories to be used in the planning processes, which consider the soil quality as a source of precious resources to be protected and enhanced.

Keywords Urbanisation · Urban landscape · Sustainability · Urban Europe

1.1 The Quest for Sustainable Cities

In the face of an increasingly growing population globally and, consequently, a demand for quality food in continuous growth, the maintenance of soil fertility and their ability to provide goods and services is a topic of particular interest in the international scientific debate (FAO 1998; Gomiero 2016; Salvati et al. 2018b). In addition to their production function, soil resources contribute to those often neglected regulatory and support services, on which not only food requirements but also other sources of psycho-physical well-being depend (Rodrigo-Comino et al. 2018); the protection and enhancement of soil resources are also generators of the cultural and aesthetic

I. Tombolini et al., *Land Quality and Sustainable Urban Forms*, Springer Geography,
https://doi.org/10.1007/978-3-030-94732-3_1

values of the landscape associated with ecosystems in a good environmental state (Brevik 2009; Jónsson and Davíðsdóttir 2016).

As it is well-known, urban sprawl competes with the availability of productive soil (Brueckner 2000; Catalán et al. 2008; Ciommi et al. 2018). According to the Thematic Strategy of the European Union for the protection of the soil elaborated by the European Commission in 2006, the covering of the soil with waterproof materials—known as soil sealing—represents one of the main soil degradation processes that occur in European countries. This phenomenon, which is continuously growing—as certified by the European Environment Agency in 2015, between 2000 and 2006 the number of sealed areas across Europe increased by about 2.7%—involves the complete loss of soil functions and is, in most cases, irreversible (Fini et al. 2017; Carlucci et al. 2018; Bimonte and Stabile 2019).

In addition, from the United Nations report "Scientific conceptual framework for land degradation neutrality" published in 2017 by Orr et al. (2017) it is noted that the conversion of natural and agricultural areas into settlements is a transition universally considered as negative for the soil, favoring its degradation (Salvati et al. 2017). It is precisely this aspect of soil degradation that we want to deal with in this book, which arises from the observation that the urban conversion of soils crosses all the expressions of landscape values in an increasingly undifferentiated way (Ciommi et al. 2019).

An issue related to this theme, probably still unexplored or not studied in depth in the scientific literature, is represented by the relationship between urban sprawl and land quality, i.e. the ability of the soil to support agricultural productivity and/or natural vegetation (Salvati et al. 2018). This question can be declined through two questions, to which in the present book attempts are made to give qualitative and quantitative answers: what type of soil is sealed by urbanization processes? How do the various forms of urban expansion (dispersed, compact or intermediate) influence land quality?

Through these questions and the resulting answers, we want to propose a reflection on the need for a change or a passage, which should probably also be reflected in planning practices, from a vision that works only on sealing toward a conception that considers the soil as a living entity, capable of hosting and supporting a varied multiplicity of organisms (Zambon et al. 2018). We, therefore, propose an expansion of interpretative categories to be used in the planning processes, which consider the soil quality as a source of precious resources to be protected and enhanced (Colantoni et al. 2016). A broader look could be decisive for structuring and choosing the most convenient strategic guidelines and territorial development policies, respecting the other elements that complete the decision-making framework.

The specific objective is to evaluate the impact that the expansion of cities, observed in its various forms, has on land quality, also providing methodological and interpretative tools to start desirable monitoring of the dynamics of this phenomenon (Ciommi et al. 2017b). The integration of soil quality into land-use planning is a relatively new concept that reflects the general commitment to sustainability and is based on awareness of the consequences of soil degradation processes. In the contemporary debate on sustainability issues—a term that indicates the increasingly urgent need not to compromise the ability of future generations to meet their needs while ensuring the economic and social development of the present generation (World Commission on Environment Development—WCED 1987)—urbanization could be considered the key to regional and global sustainability, while for others *urban sustainability*[1] is to be considered an oxymoron.

2007 was a historic moment for human civilization: we transformed ourselves from an "agricultural species" to a predominantly "urban" one. If in 1800 only 2% of the world's human population lived in urbanized areas, this percentage jumped to 14% in 1900 and 30% in 1950. In 2007, we crossed the 50% threshold and this trend shows no signs of slowing down. Urban areas have become the primary habitat for human beings: cities, more and more often, are the places where people live and it is, therefore, essential to make sustainability a reality in these same places (Wu 2010).

If to date the creation of cities has been one of the most remarkable results by humans, the development of sustainable cities could represent an even greater challenge for the future. For this reason, too, interest in urban sustainability is rapidly expanding through transdisciplinary studies, as anthropic and biophysical structures and dynamics are inextricably linked in cities (Childers et al. 2014).

1.2 Sustainable Urban Forms

Generally, urban form means a set of characteristics related to land-use models, urban transport systems and urban design (Handy 1996). Kevin Lynch (Lynch 1984) defines urban form as "the spatial model of large, inert and permanent physical objects in a city". According to this definition, the form is therefore the result of aggregations of more or less repetitive elements. But the urban form can also be considered as the result of the grouping of many elements-concepts, where the elements of the concepts, which repeat and combine to form the so-called urban models, can be road patterns, dimensions and shape of the agglomerations, the shape of roads, configuration of building plots, parks and public spaces, and so on (Jabareen 2016).

The shape of the contemporary city is sometimes associated with several environmental issues (Newman and Kenworthy 1989; Beatley et al. 1997; Alberti et al. 2007; Haughton 2016). The U.S. Environmental Protection Agency (US EPA 2014) concludes in Our Built and Natural Environment that urban form directly influences habitat, ecosystems, endangered species and water quality through soil consumption, habitat fragmentation and the spread of waterproof surfaces. Furthermore, the urban

[1] From the reading of the texts of some authors (Alberti and Susskind 1996; Spiekermann et al. 2003), it emerges that urban sustainability is basically the sustainability of the urban landscape as a whole and, as such, has a lot to do with the composition and configuration of the urban landscape that extends more and more beyond the conventional limit of the city. A sustainable city must fundamentally strike a fair balance between environmental protection, economic development and social well-being, minimizing the consumption of space and resources, optimizing the urban form to facilitate urban flows, protecting both the ecosystem and human health, ensuring equal access to resources and services and maintaining cultural and social diversity and integrity.

form affects transport, which in turn affects air quality; on the loss of agricultural land, wetlands and other natural areas; on soil pollution and its contamination; on the climate (Cervero 1998). There is therefore an increasingly pressing need to modify not only our behavior but also the design of the constructed form (Chelli et al. 2016).

The conceptual relevance, now widespread at the popular level, of "sustainable development" has rekindled the discussion on the shape of cities, motivating scholars and professionals from different disciplines to reflect on which settlement forms can meet sustainability requirements and allow better functionality to environments built. Starting especially from the 1990s, new structures have been proposed for the redesign and restructuring of urban places to achieve sustainability (Salvati et al. 2018a). These approaches have been addressed at different spatial scales: (1) at regional and metropolitan level (Forman et al. 1995; Wheeler 2016); (2) at the city level (e.g. Girardet 1999; Burton et al. 2003; Gibbs et al. 2016); (3) at the community level (e.g. Nozick 1992; Corbett et al. 1999; Paulsen 2014); and (4) at the building level (Woolley et al. 1998; Edwards and Turrent 2016; Satterthwaite 2016). A critical review of these approaches revealed a general lack of agreement on the most desirable urban form in the context of sustainability (see e.g. Frey 2003; Tomita et al. 2003; Burton et al. 2004).

The concept of sustainable development, therefore, revitalizes the debate on the urban form (e.g. Cortinovis et al. 2019), further develops existing approaches and enhances them with environmental rationalization, more precisely with the principles related to environmental sustainability and ecological design (Pili et al. 2017).

Naess (2012) identified two main models of sustainable urban development. The compact city and, on the other hand, the "green" one, which is a more open type of urban structure characterized by buildings, agricultural fields and other green areas created as a sort of a mosaic pattern (Pili et al. 2019). The main advantage of the compact city appears to be that of promoting less energy-intensive activities (Frey 2003) and implies less land consumption. However, some authors (e.g. Breheny 1992) believe that this model has little prospect for effectively reversing the more rooted tendencies of decentralization. Furthermore, the compact city may imply the refusal of suburban and semi-rural life, the abandonment of rural communities, less availability of green and open spaces, the increase in congestion and/or the increase in segregation and less autonomy at the local level (Frey 2003).

In addition, Holden (2004) stated that the definition of a compact city encompasses two different pairs of concepts that have not yet been distinguished and characterized: centralization-decentralization and concentration-sprawl (Høyer and Næss 2010). The first refers to population distribution models in national contexts, on a larger scale, the second one to development processes within urban areas (De Rosa and Salvati 2016). From the early 1960s and the advent of the automotive era, urban development can be considered as characterized by a "centralized sprawl" (Holden 2004). This means that the pattern of national population distribution and sprawl is centralized in each of the urban agglomerations (Salvati 2018). This configuration, which goes well with the concept of megacities, is opposed to that in which smaller

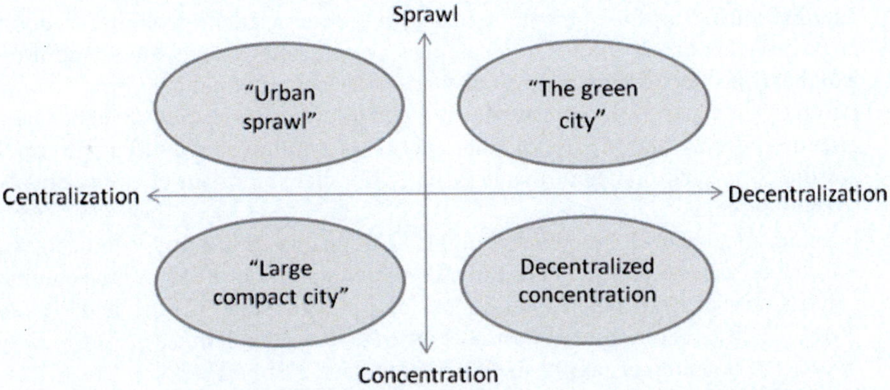

Fig. 1.1 Four possible urban form models, following the sprawl-concentration gradient, the urban scale, and the centralization-decentralization gradient at the national scale (Image modified by Holden E (2004) Ecological footprints and sustainable urban form. J Hous Built Environ 19(1):91–109)

cities and compact cities alternate (which is associated with the concept of decentralized concentration), which probably allow a more favorable ecological footprint (Fig. 1.1).

Recently, (Jabareen 2016) identified four sustainable urban forms, which present numerous overlaps in terms of ideas and concepts.

- Compact city: the distinctive concepts of the compact city are high density and compactness. Compactness also refers to urban contiguity (and connectivity), suggesting that future urban development should take place near existing urban structures (Wheeler 2016). When this concept is applied to the existing urban fabric instead of the one to be designed, it refers to the containment of further sprawl rather than the reduction of the current sprawl (Crang 2000; Wakabayashi 2002). The compactness of urban space can minimize the transport of energy, water, materials and people (Elkin 1991). Intensification, the main strategy to achieve compactness, uses urban soil more efficiently by increasing the density of settlements and activities. The intensification of the built form includes the development of previously underdeveloped urban areas, the redevelopment of existing buildings or previously developed sites, subdivisions and conversions, additions and extensions (Salvati 2014b). For many designers and scholars, compactness is the crucial type to implement in order to achieve sustainability, for four main reasons. The first is that a contained and compact city favors the protection of rural systems (McLaren 1992). The second reason is related to improving the quality of life, including social interactions and immediate access to services and facilities (Ciommi et al. 2017b). The third reason is the reduction of energy consumption, thanks to the higher density of the buildings which can favor district heating or combined systems of heat and energy; and the fourth is the reduction of greenhouse gas emissions through the reduction of transport (Colantoni et al. 2015). In addition, mixed land uses are promoted with the compact city: zoning for mixed or heterogeneous use allows to locate nearby compatible, diversified land uses, and therefore also to reduce travel distances between different activities.[2]

- Eco-city: emphasizes the spread of ecological solutions in the city, ecological and cultural diversity and the use of solar energy. Furthermore, eco-city approaches are linked to environmental management and other important environmentally friendly policies.
- Neotraditional development: highlights sustainable transport, compactness, mixed land use and greening, diversity. Diversity represents the social and cultural context of urban form. It is "a multidimensional phenomenon" (Turner and Murray 2016) which promotes further desirable urban features, including a greater variety of types of housing, of family size age, culture and incomes.
- Urban containment: claims the compactness policies. Urban containment policies include the issue of regulatory boundaries for urban growth, the delineation and design of green belts, controls on the model and density of development, the reduction of new residential development in agricultural areas, the incentive to adapt to develop new infrastructure, limiting the release of new residential permits, tax incentives and a variety of other measures (Nelson et al. 2004). In general, urban containment policies seek to use at least three different types of tools to model metropolitan growth (Salvati and Lamonica 2020): green belts and urban growth boundaries are used to influence push factors (to limit attraction to the outside) and urban services to influence pull factors (to increase attraction toward the interior).

There are probably more sustainable urban forms, but, in general, all have a high density and adequate diversity. They are not dispersive and have mixed uses of the soil, which are designed to favor sustainable transport and the spread of ecological solutions (Salvati 2014a). To conclude, we state that sustainable urban forms aim to achieve different objectives. The most important ones are the reduction of energy consumption, wastes, pollution, the use of the car and the conservation of open spaces and sensitive ecosystems, protecting of more livable and oriented environments to the community.

1.3 From *Land* to *Landscape*

Land, a term that is often considered by several languages as soi or surface, is inextricably linked to the concept of *landscape*. The terms "land" and "soil" are not exactly interchangeable but, conceptually, the first includes the second, to the point that in the 12th Conference of the United Nations Convention (UNCCD) held

[2] As argued by Newman (1997), in recent decades, urban planning has instead led to the "breakdown" of cities through the use of a rigid zoning that separates individual land uses. The result was to produce a city with less local diversity and more traffic, as well as less safety and less appeal of local roads.

in 2015, the parties agreed on a differentiation between the two terms. While the word "soil" indicates one of the most important natural resources for our planet, the term "land" indicates an ecologically multifunctional system, whose natural capital, consisting of the soil and the biodiversity associated with it, interacting with the water and the atmosphere, generates a flow of ecosystem services that support human well-being, ensuring the life and sustenance of individuals and communities, animals and plants (UNCCD 2015; paragraph 22). The "land" system or "lands" also constitute a fundamental capital for the development of most economic activities: soils of high quality, that is, fertile, have ensured the maintenance of vital agriculture for millennia and shown great resilience skills. The recent phenomena of urban transformation are, however, causing socio-economic and environmental changes which in the long term can have unpredictable impacts, especially in ecologically fragile areas (Incerti et al. 2007; Cerdà et al. 2010; Cowie et al. 2018).

As defined in even more detail in the Thematic Strategy for Soil Protection, the soil is the top layer of the earth's crust, consisting of mineral particles, organic matter, water, air and living organisms, which represents the interface between earth, air and water and hosts much of the biosphere (Kosmas et al. 2016). Given the extremely long times of soil formation, it can be considered that it is a substantially non-renewable resource (Karlen et al. 1997; Soil Survey Staff 2014): "*Soil is one of humanity's precious assets. It allows the life of plants, animals and humans on the surface of the earth*[3]".

Good quality soil is capable of ensuring many ecological, economic and social functions (Carter 2002; Bastida et al. 2008; Karamesouti et al. 2015). Resuming results of a long tradition in soil science, these functions can be traced back to seven main points.

- Filter, buffer capacity and transformation of various materials and substances. The soil plays an important protective function, through a filter and barrier action, which allows mitigating the effects of polluting substances, hindering their passage into groundwater or the food chain. Its role in the water cycle is crucial. It can control the deep transport of solutes and the flow of water on the surface, to regulate the absorption by vegetation and to create favorable conditions for the degradation of polluting substances. The protective value of soil coverings depends on the physical–mechanical properties of the soil, which determine a filter and barrier action against the movement of pollutants, their chemical-physical properties, which determine their buffering capacity, and biological activity, which allows the biochemical and microbiological decomposition of substances released into the soil.
- Carbon stock. Soil organic carbon constitutes the set of organic compounds deriving from living material or which has been in the past, with the exclusion of only the living vegetable biomass, which are found in the soil or on its surface. In terms of carbon storage quantities, the oceans constitute the prevailing reservoir but in terms of importance, the soil-vegetation reservoir is the one that

[3] Art. 1, European Soil Charter, Council of Europe, 1972.

prevails because even if the quantities are smaller (about 6% of that of the oceans) the variations and the exchanges carbon are faster and the total balance can be directly influenced by human action (fire control, erosion control, maintenance or reconstitution of forest cover, agronomic management).

- Biodiversity pool and genetic reserve. The soil is a very complex environment that functions as a habitat for a very high number of organisms concentrated mainly in the first centimeters from the surface. In the intricate three-dimensional matrix of the soil, these organisms interact with each other in a dense food network, creating a complex system of biological activities. Some of these live permanently within the soil, others pass only some stages of their biological cycle, or use it as protection in times of difficulty and stasis (cysts, pupation, wintering, summer).
- Biomass production. Agricultural and food production, essential for human survival, and forestry depend entirely on the soil. Almost all vegetation, including pastures, crops and forests, insist on the soil which, in addition to physical support, provides plants with water and nutrients.
- Procurement of mineral resources and raw materials. Almost all human works require the use of raw materials such as stone and incoherent rocks, metal and non-metallic minerals, natural gas, hydrocarbons, fuels and waters. Soil is the means of access to these raw materials whose possession is considered a strategic factor for the economic development of a country and the evolution of international relations. However, the impacts on soil generated by quarries and mines are manifold; they begin during the extraction activity and continue over time with the activity completed, especially where an adequate environmental restoration plan is not envisaged.
- A platform for human activities. Soil is one of the key elements of human-populated ecosystems as it supports several important activities taking place there. The term "urban soil" is often used, a generalization that is used to mean any soil, natural, modified or created by human, that exists in an urban or industrial area. In these areas, the soil performs a series of functions which overlap traditionally recognized functions, such as aesthetic, recreational (gardens, tree-lined avenues, public parks), study and conservation of cultural heritage.
- Heritage. The soil, as part of the territory in which it is inserted, has always been analyzed and studied about its economic interest represented by its primary qualities, attributable to its fertility and its agricultural productivity. However, in recent years other aspects, previously inadequately recognized, have also been evaluated, which consider the soil an important place and means of preserving a series of values that are in all respects comparable to the "cultural asset". From this point of view, the soil is of considerable importance and cultural interest as it can preserve important geo-paleontological, archaeological and/or geo-archaeological evidence.

Taking into account the importance of the numerous functions that the soil performs, if the open territory, still undeveloped, has until recently been considered the neutral background of the city or as a reserve for building expansion, it must be seen today in new terms: it is no longer possible to think of the territory in a fragmented way, as has often been done in recent decades, but it is necessary to think about its complexity, its formation and the continuous breakages that arise from it, considering together the articulation of the spaces, the plurality of their uses and the interaction of their functions through an appropriate qualification of the same. Italian architect Bernardo Secchi had raised an important question: over the years the design of the urban planners and, in particular, their way of cartographically representing their project has changed as in a sort of return to the origins, having increased its iconic and metaphorical character (Secchi and Vigano 2011). Above all, in the urban project, this fundamental resource must be understood in its thickness of support, of an infrastructure that supports artificial settlements such as the affirmation of production processes and the deployment of the material and immaterial networks that make up the landscape.

The link between the idea of land and that of landscape exists historically. According to Jackson (1986), the term land was originally connected to the agricultural use of the land, indicating "the plot of land or the furrows of a field that were rotated annually". It indicated a defined space and was associated with farmers with some form of spatial measurement. Jackson also argued that the word *landscape*, in ancient English or the Anglo-Saxon meaning, probably indicates a set of land or more generally a system of agricultural spaces, or the vernacular landscape composed of the three essential components: villages, arable lands and pastures. However, the landscape also has a cultural connotation, referring to human's vision of the land (Fig. 1.2).

> Land becomes landscape when seen by man, revealing the record of his activities on the surface of the earth and his relationship with his environment. The perception of landscape reveals his attitude towards it and generates emotions ranging from distrust and fear to reassurance and delight. These may arise from the view of a real landscape or the imagery of poet, painter or writer.

From the aforementioned words of Hunter,[4] author of the book "Land into Landscape", it appears that the nuances of the term landscape or landscape have both a territorial and spatial meaning and one more related to aesthetics. Lörzing (2001), author of the book "The nature of landscape: a personal quest", noted that this duality of meaning appears in various languages and can be compared to a more technical

---

[4] In his first book, 1985, "Land into landscape", John Michael Hunter, born in Edinburgh in 1932, told about his approach to the landscape, his perception of human's relationship with it, and some account of what he himself he had lived. He worked as a planner for Essex County making a significant contribution to environmental management and conservation. In 1971, in particular, he entered the council of the county planning department, in a period of growing concern for environmental issues: the countryside landscape had undergone a considerable transformation, becoming almost bare, due to the removal of the hedges and the spread of intensive breeding. Following a conference entitled "Landscape in Decline", in 1972, Hunter was instrumental in creating a landscape conservation program, thanks to which grants arose to provide advice and assistance for the restoration of the tree cover of the countryside. This laid the foundations for an approach to the management of agriculture and the countryside that has been widely imitated and which is fundamental for the policies pursued today by the current English ministry for the environment and agricultural policies (Hunter 1999).

Fig. 1.2 Hora, a characteristic village dating back to 1200 AD, of Folegandros, a small island in the Cyclades archipelago, seen from Panaghia (photo by Ilaria Tombolini, September 2017)

"objective" vision on the one hand or to a more vision personal, emotional "subjective" on the other one. The meaning of landscape, in a broad sense, is enshrined in the text of the *European Landscape Convention* where it is defined as follows:

> [...] an area, as perceived by people, whose character is the result of the action and interaction of natural and/or human factors. The term 'landscape' is thus defined as a zone or area as perceived by local people or visitors, whose visual features and character are the result of the action of natural and/or cultural (that is, human) factors. This definition reflects the idea that landscapes evolved through time, as a result of being acted upon by natural forces and human beings. It also underlines that landscape forms wholly natural and cultural components are taken together not separately.[5]

Article 2 of the Convention also specified that it applies to the entire territory of the Parties and covers natural, rural, urban and peri-urban areas. "*[...] it includes land, inland, water and marine areas. It concerns landscapes that might be considered outstanding as well as every day or degraded landscape*".

The landscape is therefore not a mere aesthetic context to be preserved but represents a characteristic element of the environmental and cultural identity of each community (Swanwick 2009), and as such a heritage to be handed down to future generations. The soil, which it can consider as the "living skin" of the planet Earth

[5] European Council, 2000.

(Dent et al. 2005) or earth's epidermis (Tricart 1962), for its characteristics of the finite and non-renewable resource must be recognized as a common good whose availability circumscribes the physical space within which each community establishes and hands down its bases of subsistence, well-being, freedom and cultural identity.

1.4 Green Policies and Quality of Life

Recent socio-economic dynamics leave us more fragile cities, massively influenced by external shocks, and justify a partial rethinking of spatial policies and developmental measures in local communities—both inner cities and suburbs—exposed to aging and poverty, and less attractive for immigrants from abroad. Being required to provide differentiated services for a declining population, the new European policies—largely inspiring to the "Green Deal" framework—should reconcile demographic dynamics and the territorial specificity of local contexts, promoting appropriate spatial infrastructures in metropolitan areas and containing urban sprawl as inappropriate way of metropolitan growth. In this perspective, a comparative analysis of local-scale urban dynamics contributes to a more comprehensive understanding of the latent reorganization of metropolitan spaces with sequential economic downturns, from expansion to recession. In line with the "Green Deal" perspectives, strategies coping with the sustainable development of large Mediterranean cities should create the appropriate conditions for equity, cohesion, competitiveness, and environmental security on both regional and metropolitan scales. Place-specific policies and multi-scale planning considering together competitiveness, sustainability, and resilience targets are increasingly requested to delineate new developmental paths for the coming cities, in the Mediterranean basin and, more generally, in Europe.

Resilience characteristic of local systems should be reinforced with an appropriate governance of their "socio-demographic engine". Being unable to engage global competition dynamics, cities with stagnant economies should recognize the legacy of traditional settlement forms, abandoning the long experience of exurban development, and putting "quality of life" at the core of the normative debate on sustainable urbanization. Quality of life can be improved exploiting best technological practices. Smart (and compact) cities represent an opportunity for the intrinsic recovery of degraded and declining settlements. In a context of limited spending capacity of national and regional governments, the European post-pandemic cohesion and recovery funds can significantly contribute in this direction.

Recent socio-economic dynamics have not only had negative implications on metropolitan growth, instead representing a natural process of urban de-concentration in hyper-compact contexts. This process offers some unexpected opportunities for a truly sustainable, integrated and spatially balanced (post-crisis) development. In this perspective, how much the actual pandemic crisis can drive city decline should be debated more extensively. We believe that, the pandemic crisis could represent a further driver of urban decline if not properly managed. At the same time, the opportunities for urban–rural re-balance inherent in a more intense phase of counter-urbanization, should stimulate more coordinated actions of regeneration and recovery of old spaces, redevelopment of vacant land, regulation of smart-working, and moderate settlement densification compatible with the demand of an increasingly volatile real estate market.

References

Alberti M, Susskind L (1996) Managing urban sustainability: an introduction to the special issue. Environ Impact Assess Rev 16:213–221. https://doi.org/10.1016/S0195-9255(96)00070-4

Alberti M, Booth D, Hill K, Coburn B, Avolio C, Coe S, Spirandelli D (2007) The impact of urban patterns on aquatic ecosystems: an empirical analysis in Puget lowland sub-basins. Landscape Urban Plan 80:345–361. https://doi.org/10.1016/j.landurbplan.2006.08.001

Bastida F, Zsolnay A, Hernández T, García C (2008) Past, present and future of soil quality indices: a biological perspective. Geoderma 147:159–171. https://doi.org/10.1016/j.geoderma.2008.08.007

Beatley T, Timothy B, Beatley T, Manning K (1997) The ecology of place: planning for environment, economy, and community. Island Press

Bimonte S, Stabile A (2019) The effect of growth and corruption on soil sealing in Italy: a regional environmental Kuznets Curve Analysis. Environ Resour Econ 74:1497–1518. https://doi.org/10.1007/s10640-019-00376-1

Breheny M (1992) Sustainable development and urban form: an introduction. In: Sustainable development and urban form. Pion, London, UK

Brevik EC (2009) Soil, food security, and human health. In: Verheye W (ed) Soils, Plant Growth and Crop Production, Encyclopedia of Life Support Systems (EOLSS), Developed under the Auspices of the UNESCO. EOLSS Publishers, Oxford, UK

Brueckner JK (2000) Urban Sprawl: Diagnosis and Remedies. Int Reg Sci Rev 23:160–171. https://doi.org/10.1177/016001700761012710

Burton E, Jenks M, Williams K (2003) Achieving the compact city through intensification: an acceptable option? The compact city. Routledge, London, UK. https://doi.org/10.4324/9780203362372-15

Burton E, Jenks M, Williams K (2004) The compact city: a sustainable urban form? Taylor & Francis

Carlucci M, Chelli FM, Salvati L (2018) Toward a new cycle: short-term population dynamics, gentrification, and re-urbanization of Milan (Italy). Sustainability 10:1–12

Carter MR (2002) Soil quality for sustainable land management. Agron J 94:38–47

Catalán B, Saurí D, Serra P (2008) Urban sprawl in the Mediterranean? Patterns of growth and change in the Barcelona Metropolitan Region 1993–2000. Landscape Urban Plan 85:174–184

Cerdà A, Lavee H, Romero-Díaz A, Hooke J, Montanarella L (2010) Preface: soil erosion and degradation in Mediterranean type ecosystems. Land Degrad Dev 21:71–74. https://doi.org/10.1002/ldr.968

Cervero R (1998) The transit metropolis: a global inquiry. Island Press, Washington

Chelli FM, Ciommi M, Emili A, Gigliarano C, Taralli S (2016) Assessing the equitable and sustainable well-being of the Italian Provinces. Int J Uncertainty Fuzziness Knowl-Based Syst 24:39–62. https://doi.org/10.1142/S0218488516400031

Childers DL, Steward TA, Pickett JM, Grove LO, Whitmer A (2014) Advancing urban sustainability theory and action: challenges and opportunities. Landscape Urban Plan 125:320–328. https://doi.org/10.1016/j.landurbplan.2014.01.022

Ciommi M, Chelli FM, Carlucci M, Salvati L (2018) Urban growth and demographic dynamics in Southern Europe: toward a new statistical approach to regional science. Sustainability 10:2765. https://doi.org/10.3390/su10082765

Ciommi M, Chelli FM, Salvati L (2019) Integrating parametric and non-parametric multivariate analysis of urban growth and commuting patterns in a European metropolitan area. Qual Quant 53:957–979. https://doi.org/10.1007/s11135-018-0798-2

Ciommi M, Gentili A, Ermini B, Gigliarano C, Chelli FM, Gallegati M (2017a) Have your cake and eat it too: the well-being of the Italians (1861–2011). Soc Indic Res 134:473–509. https://doi.org/10.1007/s11205-016-1450-y

Ciommi M, Gigliarano C, Emili A, Taralli S, Chelli FM (2017b) A new class of composite indicators for measuring well-being at the local level: an application to the Equitable and Sustainable Well-being (BES) of the Italian Provinces. Ecol Ind 76:281–296. https://doi.org/10.1016/j.ecolind.2016.12.050

Colantoni A, Grigoriadis E, Sateriano A, Venanzoni G, Salvati L (2016) Cities as selective land predators? A lesson on urban growth, deregulated planning and sprawl containment. Sci Total Environ 545–546:329–339. https://doi.org/10.1016/j.scitotenv.2015.11.170

Colantoni A, Mavrakis A, Sorgi T, Salvati L (2015) Towards a 'polycentric' landscape? Reconnecting fragments into an integrated network of coastal forests in Rome. Rendiconti Lincei 26:615–624. https://doi.org/10.1007/s12210-015-0394-5

Corbett M, Corbett J, Thayer RL (1999) Designing sustainable communities: learning from village homes. Island Press, Washington, DC

Cortinovis C, Haase D, Zanon B, Geneletti D (2019) Is urban spatial development on the right track? Comparing strategies and trends in the European Union. Landscape Urban Plan 181:22–37. https://doi.org/10.1016/j.landurbplan.2018.09.007

Cowie AL, Orr BJ, Castillo Sanchez VM, Chasek P, Crossman ND, Erlewein A, Louwagie G et al (2018) Land in balance: the scientific conceptual framework for Land Degradation Neutrality. Environ Sci Policy 79:25–35. https://doi.org/10.1016/j.envsci.2017.10.011

Crang M (2000) Public space, urban space and electronic space: would the real city please stand up? Urban Stud 37:301–317. JSTOR

De Rosa S, Salvati L (2016) Beyond a 'side street story'? Naples from spontaneous centrality to entropic polycentricism, towards a 'crisis city.' Cities 51. Current Research on Cities: 74–83. https://doi.org/10.1016/j.cities.2015.11.025

Dent DL, Hartemink AE, Kimble J (2005) Soil—Earth's living skin. Planet Earth-IUGS-UNESCO, Trondheim

Edwards B, Turrent D (2016) Sustainable housing: principles and practice. CRC Press

FAO (1998) Topsoil characterization for sustainable land management. Food and Agriculture organization of the United Nations Land and water development division soil resources, management and conservation service, pp 1–49

Fini A, Frangi P, Mori J, Donzelli D, Ferrini F (2017) Nature based solutions to mitigate soil sealing in urban areas: Results from a 4-year study comparing permeable, porous, and impermeable pavements. Environ Res 156:443–454. https://doi.org/10.1016/j.envres.2017.03.032

Forman RT, Forman RTT (1995) Land mosaics: the ecology of landscapes and regions. Cambridge University Press

Frey H (2003) Designing the city: towards a more sustainable urban form. Taylor & Francis, Rochester, NY

Gibbs DC, Longhurst J, Braithwaite C (2016) 'Struggling with sustainability': weak and strong interpretations of sustainable development within local authority policy. Environ Planning A. https://doi.org/10.1068/a301351

Girardet H (1999) Creating sustainable cities: 02. Totnes, Devon

Gomiero T (2016) Soil degradation, land scarcity and food security: reviewing a complex challenge. Sustainability 8:281. https://doi.org/10.3390/su8030281

Handy S (1996) Methodologies for exploring the link between urban form and travel behavior. Transp Res Part D: Transp Environ 1:151–165. https://doi.org/10.1016/S1361-9209(96)00010-7

Haughton G (2016) Environmental justice and the sustainable city. J Planning Educ Res. https://doi.org/10.1177/0739456X9901800305

Holden E (2004) Ecological footprints and sustainable urban form. J Hous Built Environ 19:91–109. https://doi.org/10.1023/B:JOHO.0000017708.98013.cb

Høyer KG, Næss P (2010) The ecological traces of growth: economic growth, liberalization, increased consumption—and sustainable urban development? J Environ Policy Planning World. https://doi.org/10.1002/jepp.84

Hunter JM (1999) The Essex landscape: a study of its Form and history. Essex Record Office Publications, Chelmsford, UK

Incerti G, Feoli E, Salvati L, Brunetti A, Giovacchini A (2007) Analysis of bioclimatic time series and their neural network-based classification to characterise drought risk patterns in South Italy. Int J Biometeorol 51:253–263. https://doi.org/10.1007/s00484-006-0071-6

Jabareen YR (2016) Sustainable urban forms: their typologies, models, and concepts. *J Planning Educ Res*. https://doi.org/10.1177/0739456X05285119

Jackson JB (1986) Discovering the Vernacular Landscape. Londra, New Haven

Jónsson JÖG, Davíðsdóttir B (2016) Classification and valuation of soil ecosystem services. Agric Syst 145:24–38

Karamesouti M, Detsis V, Kounalaki A, Vasiliou P, Salvati L, Kosmas C (2015) Land-use and land degradation processes affecting soil resources: evidence from a traditional Mediterranean cropland (Greece). CATENA 132:45–55. https://doi.org/10.1016/j.catena.2015.04.010

Karlen DL, Mausbach MJ, Doran JW, Cline RG, Harris RF, Schuman GE (1997) Soil quality: a concept, definition, and framework for evaluation (a guest editorial). Soil Sci Soc Am J 61:4–10. https://doi.org/10.2136/sssaj1997.03615995006100010001x

Kosmas C, Karamesouti M, Kounalaki K, Detsis V, Vassiliou P, Salvati L (2016) Land degradation and long-term changes in agro-pastoral systems: an empirical analysis of ecological resilience in Asteroussia—Crete (Greece). CATENA 147:196–204. https://doi.org/10.1016/j.catena.2016.07.018

Lörzing H (2001) The nature of landscape: a personal quest. 010 Publishers

Lynch K (1984) Good city form. MIT Press

Elkin T, McLaren D, Mayer H (1991) Reviving the City: Towards Sustainable Urban Development. Friends of the Earth, London

Mclaren D (1992) Compact or Dispersed? Dilution is No Solution. Built Environ (1978-) 18:268–284. JSTOR

Naess P (2012) Urban form and travel behavior: experience from a Nordic context. J Transport and Land Use 5. https://doi.org/10.5198/jtlu.v5i2.314

Nelson AC, Burby RJ, Feser E, Dawkins CJ, Malizia EE, Quercia R (2004) Urban containment and central-city revitalization. J Am Plann Assoc 70:411–425. https://doi.org/10.1080/01944360408976391

Newman PWG (1997) Greening the city: The ecological and human dimensions of the city can be part of town planning. In: Roseland M (ed) Eco-city dimensions: Healthy communities, healthy planet, pp 14–24. Gabriola Island, British Columbia, Canada: New Society Publishers.

Newman PWG, Kenworthy JR (1989) Gasoline consumption and cities. J Am Plann Assoc 55:24–37. https://doi.org/10.1080/01944368908975398

Nozick M (1992) No place like home: building sustainable communities. Canadian Council on Social Development, Ottawa

Orr BJ., Cowie AL, Castillo Sanchez VM, Chasek P, Crossman ND, Erlewein A, Louwagie G, et al (2017) Scientific conceptual framework for land degradation neutrality, 1st edn. United Nations Convention to Combat Desertification (UNCC), Bonn, Germany

Paulsen K (2014) Geography, policy or market? New evidence on the measurement and causes of sprawl (and infill) in US metropolitan regions. Urban Stud 51:2629–2645. https://doi.org/10.1177/0042098013512874

Pili S, Grigoriadis E, Carlucci M, Clemente M, Salvati L (2017) Towards sustainable growth? A multi-criteria assessment of (changing) urban forms. Ecol Ind 76:71–80. https://doi.org/10.1016/j.ecolind.2017.01.008

Pili S, Serra P, Salvati L (2019) Landscape and the city: Agro-forest systems, land fragmentation and the ecological network in Rome, Italy. Urban For Urban Greening 41:230–237. https://doi.org/10.1016/j.ufug.2019.04.016

Rodrigo-Comino J, Senciales JM, Cerdà A, Brevik EC (2018) The multidisciplinary origin of soil geography: a review. Earth Sci Rev 177:114–123. https://doi.org/10.1016/j.earscirev.2017.11.008

Salvati L (2014a) Towards a polycentric region? The socio-economic trajectory of Rome, an 'Eternally Mediterranean' City. Tijdschr Econ Soc Geogr 105:268–284. https://doi.org/10.1111/tesg.12054

Salvati L (2014b) Agro-forest landscape and the 'fringe' city: a multivariate assessment of land-use changes in a sprawling region and implications for planning. Sci Total Environ 490:715–723. https://doi.org/10.1016/j.scitotenv.2014.05.080

Salvati L (2018) Population growth and the economic crisis: understanding latent patterns of change in Greece, 2002–2016. Lett Spat Resour Sci 11:105–126. https://doi.org/10.1007/s12076-018-0204-7

Salvati L, Lamonica GR (2020) Containing urban expansion: densification vs Greenfield development, socio-demographic transformations and the economic crisis in a Southern European City, 2006–2015. Ecol Ind 110:105923. https://doi.org/10.1016/j.ecolind.2019.105923

Salvati L, Carlucci M, Grigoriadis E, Chelli FM (2018a) Uneven dispersion or adaptive polycentrism? Urban expansion, population dynamics and employment growth in an 'ordinary' city. Rev Reg Res 38:1–25. https://doi.org/10.1007/s10037-017-0115-x

Salvati L, Ferrara A, Chelli F (2018b) Long-term growth and metropolitan spatial structures: an analysis of factors influencing urban patch size under different economic cycles. Geogr Tidsskr-Danish J Geogr 118:56–71. https://doi.org/10.1080/00167223.2017.1386582

Salvati L, Guandalini A, Carlucci M, Chelli FM (2017) An empirical assessment of human development through remote sensing: evidences from Italy. Ecol Ind 78:167–172. https://doi.org/10.1016/j.ecolind.2017.03.014

Salvati L, Zambon I, Chelli FM, Serra P (2018c) Do spatial patterns of urbanization and land consumption reflect different socioeconomic contexts in Europe? Sci Total Environ 625:722–730. https://doi.org/10.1016/j.scitotenv.2017.12.341

Satterthwaite D (2016) Sustainable cities or cities that contribute to sustainable development? Urban Stud. https://doi.org/10.1080/0042098975394

Secchi, and Vigano (2011) La Ville Poreuse: Un Projet Pour le Grand Paris et La... MētisPresses, Genève

Soil Survey Staff (2014) Keys to Soil Taxonomy, 12th edn. USDA-Natural Resources Conservation Service, Washington, DC

Spiekermann K, Wegener M, Urban W (2003) Modelling urban sustainability. Int J Urban Sci 7:47–64

Swanwick C (2009) Society's attitudes to and preferences for land and landscape. Land Use Policy 26. Land Use Futures: S62–S75. https://doi.org/10.1016/j.landusepol.2009.08.025

Tomita Y, Terashima D, Hammad A, Hayashi Y (2003) Backcast analysis for realizing sustainable urban form in Nagoya. Built Environ (1978–) 29:16–24. JSTOR

Tricart J (1962) L'Epiderme de la Terre. Esquisse d'une géomorphologie appliquée. Masson, Paris

Turner RS, Murray MS (2016) Managing Growth In A Climate Of Urban Diversity: South Florida's Eastward Ho! Initiative. J Planning Educ Res. https://doi.org/10.1177/0739456X0102000304

UNCCD (2015) Report of the Conference of the Parties on its twelfth session. ICCD/COP(12)/20/Add.1. Paris, France

US EPA, OP (2014) Our Built and Natural Environments. Reports and Assessments EPA 231-R-01-002. USA

Wakabayashi M (2002) Urban space and cyberspace: urban environment in the age of media and information technology. Int J Jpn Sociol 11:6–18. https://doi.org/10.1111/1475-6781.00014

Wheeler SM (2016) Planning for metropolitan sustainability. J Planning Educ Res. https://doi.org/10.1177/0739456X0002000201

Woolley T, Kimmins S, Harrison R, Harrison P (1998) Green building handbook: volume 1: a guide to building products and their impact on the environment. E & FN Spon, London

Wu J (2010) Urban sustainability: an inevitable goal of landscape research. Landscape Ecol 25:1–4. https://doi.org/10.1007/s10980-009-9444-7

Zambon I, Benedetti A, Ferrara C, Salvati L (2018) Soil matters? A multivariate analysis of socioeconomic constraints to urban expansion in Mediterranean Europe. Ecol Econ 146:173–183. https://doi.org/10.1016/j.ecolecon.2017.10.015

Chapter 2
Toward a Sustainable Use of Land: Urbanization, Policies and (Mis)Understanding of Degradation Processes

Abstract The structure and composition of the landscape continuously evolve in space and time, influencing the physical, chemical and biological processes of the soil. These influences contribute significantly to the complex interactions between the natural environment and anthropic activities, shaping the characteristics and properties of the lands in various ways (the spatial diversification of the properties of the soil and its compaction are typical examples). In this chapter, land degradation and land quality concepts will be defined and discussed considering the issues that processes represent a threat to the sustainability and production capacity of agricultural activity. In a context in which the economy, society and the environment tend to become increasingly integrated and interconnected dimensions, issues related to sustainable development are becoming increasingly important. For this reason, we also discuss the different policies emitted by the European Union on soil protection. Soil is not subject to a complete and coherent set of rules in the Union. Existing EU policies in areas such as agriculture, water, waste, chemicals and the prevention of industrial pollution contribute indirectly to soil protection. The European Union's agenda for research and innovation policy on nature-based solutions and the renaturalization of cities aims to position the EU as a leader in the process of "innovation with nature", for more sustainable societies and resilient. Also, the concept of "Nature-Based Solutions" (NBS), which is relatively new and has been introduced to promote nature as a means of providing solutions in mitigation measures and adaptation challenges to climate change was defined. Finally, the case of Italy was analyzed to demonstrate how land degradation processes and land quality can be estimated.

Keywords Land policies · European Union · Land quality · Land degradation · Soil sealing

2.1 International and European Policies

As implicitly announced in the text of the European Landscape Convention reported in the introduction, on a global level, the lands and, consequently, the landscape are threatened by degradation processes that derive from the synergistic action of

I. Tombolini et al., *Land Quality and Sustainable Urban Forms*, Springer Geography,
https://doi.org/10.1007/978-3-030-94732-3_2

biophysical factors, often connected to climatic variations and anthropic activities. The protection and sustainable and systematic management of the soil, therefore, represent a necessity that can be no longer postponed given the diversified degradation situations that emerge from various territorial areas, both at European and international level (Kairis et al. 2013, 2015; Gonzalez-Roglich et al. 2019; Panagos and Katsoyiannis 2019).

Europe and the United Nations promote the protection of the soil and the recognition of the heritage associated with this important resource through the following objectives:

- zero net land consumption by 2050[1];
- adequately protect the soil also with the adoption of soil-related objectives as an essential resource of natural capital by 2020[2];
- align consumption with real population growth by 2030[3];
- do not increase land degradation by 2030.[4]

The definition of the targets listed above derives from a path started in 2002 when the European Commission released a "Communication" entitled "Towards a thematic strategy for soil protection" which highlighted the importance of the soil as a vital and fundamental resource not renewable, subjected to increasing pressure. The text represented for the Commission a political commitment for the protection of the soil, with the awareness of the complexity of the topic and the need for long times for the formulation of an integrated European policy capable of stopping degradation processes and effectively protecting this fundamental environmental resource (Delfanti et al. 2016; Recanatesi et al. 2016; Zambon et al. 2018).

At a European level, environmental issues have often been used to define "thematic strategies" made binding by specific Directives and aimed at establishing cooperation measures and guidelines aimed at Member States and local authorities. Thus, also in the case of the soil, in September 2006, a new Directive of the European Parliament and the Council was proposed aimed at defining an overall framework for soil protection and adoption of a thematic strategy for their protection and use: Sustainable soil.[5] This strategy placed the emphasis on preventing further soil degradation and maintaining its functions, emphasizing the need to implement good practices to reduce the negative effects of the spread of artificial, waterproof coverings on the soil (Luca Salvati et al. 2016).

The importance of good management of the territory and, in particular, of the soils was then reaffirmed by the Commission in 2011 with the Roadmap toward an efficient

---

[1] European Parliament and Council (2013, December 28) Decision no. 1386/2013 / EU of the European Parliament and of the Council of 20 November 2013 on a general Union action program on the environment up to 2020 "Living well within the limits of our planet". OJEU, L 354:171–200.

[2] Ibid.

[3] UN (2015) Transforming our World: The 2030 Agenda for Sustainable Development. A/RES/70/1, United Nations.

[4] Ibid.

[5] European Commission (2006, September 22) Thematic Strategy for Soil Protection. COM (2006) 231. Brussels.

Europe in the use of resources.[6] It was also linked to the 2020 Strategy, with the aim of an increase in the net occupation of zero lands to be achieved in Europe by 2050. Objective reaffirmed later with the approval of the Seventh Environmental Action Program, called "Living well within the limits of our planet", which also required that, by 2020, Union policies took into account their direct and indirect impacts on land use (Smiraglia et al. 2016). From a formal point of view, it is important to underline that the Seventh Environmental Program of the European Union, initialed on 20 November 2013 and entered into force in January 2014, is a Decision of the European Parliament and the Council and, therefore, has a normative nature, unlike of the Commission's 2011 roadmap, which merely outlined important political priorities. The final objective of the Program, also referring to the conclusions of the United Nations Conference on Sustainable Development held in Rio de Janeiro in 2012, concerns the wider global challenges to achieve the objectives of sustainable development with the participation of all international partners. In particular, the conclusions of the 2012 United Nations Conference on Sustainable Development made it possible to bring the issue of protection, conservation and improvement of natural resources, including soil, to public attention (Kosmas et al. 2016). The final report, "The future we want"[7] invited national governments to intervene to ensure greater attention to decisions relating to land use, at all levels of relevance, take due account of the environmental, social and economic impacts that generate degradation of lands. In addition, the importance of reversing these processes was explicitly stated and achieving, as will be seen in paragraph 2.1.2, the goal of a "land degradation neutral world".

The European Commission had already found it useful to indicate the priorities for action and the guidelines to be followed to achieve the goal of zero net land occupation by 2050 by publishing, in 2012, the guidelines to limit, mitigate and compensate the soil sealing.[8] The proposed approach was to implement policies and actions aimed at limiting, mitigating and compensating for soil sealing, to be defined in detail in the Member States and to be implemented at the national, regional and local level. In other words, the Member States should, as a priority, ensure the limitation of the waterproofing through the reduction of the conversion and transformation rate of the agricultural and natural territory and the reuse of the already urbanized areas, with the definition of realistic targets for land use at the national and regional level. If soil loss is inevitable, mitigation measures should be provided, aimed at maintaining the main functions of the soil and reducing negative effects on the environment (Luca Salvati and Zitti 2009; Luca Salvati and Carlucci 2011; Luca Salvati 2014b). Finally, all the inevitable interventions of new waterproofing of the soil should be compensated by ensuring, for example, a renaturalization of already waterproofed land or, as a last

[6] European Commission (2011, September 20) Roadmap towards an efficient Europe in the use of resources. COM (2011) 571. Brussels.

[7] UN (2012) The Future We Want. A/RES/66/288. United Nations.

[8] European Commission (2012, May 15) Guidelines on good practices to limit, mitigate and compensate for soil waterproofing. Brussels, SWD (2012) 101.

resort, in the form of economic fees, provided that they are tied to the use in soil restoration actions (Bajocco et al. 2011).

Unfortunately, the Thematic Strategy for Soil Protection (COM (2006) 231) and the relative proposal for a Directive (Soil Framework Directive (COM (2006) 232), after almost 10 years from their adoption, were withdrawn in May 2014 due to strong opposition from some Member States for reasons related mainly to subsidiarity, costs deemed an excessive and administrative burden.

2.1.1 European Union Policies for Soil Protection

Although the importance of soil and ecosystem services that it can provide is now globally recognized, at the moment, only a few Member States of the European Union have specific legislation on soil protection. Soil is not subject to a complete and coherent set of rules in the Union. Existing EU policies in areas such as agriculture, water, waste, chemicals and the prevention of industrial pollution contribute indirectly to soil protection (Luca Salvati et al. 2011). But since these policies have other objectives and fields of action, they are not sufficient to guarantee an adequate level of protection for all soils in Europe as several scientific groups are demonstrating year by year (e.g.Boellstorff and Benito 2005; Vanwalleghem et al. 2011; Rodrigo-Comino 2018).

In an interesting and detailed report published by the Berlin Ecological Institute,[9] the various legislative instruments have been identified that at European Union level concern soil protection and which based on the main topic considered are grouped into clusters. Being probably the most complete work done so far in this area, it is taken as a reference below to address the issue of soil policies at European Union level, developing the discussion around the clusters identified (Table 2.1). The degradation phenomena examined and considered useful for this research are attributable to the biophysical alterations of the soil (and non-chemical), which is why the aspects of policies focused on soil pollution and contamination are not studied in depth (Colantoni et al. 2015).

2.1.1.1 General Instruments Cluster

The policies of this cluster can be divided into three types:

- Providing long-term guidelines for certain environmental problems and human health (the Seventh environmental action program);
- Dealing with economic issues and the use of environmental resources (the Roadmap toward an efficient Europe in the use of resources and the Action Plan for the circular economy);

[9] Frelih-Larsen et al. (2016) Updated Inventory and Assessment of Soil Protection Policy Instruments in EU Member States. Final Report to DG Environment. Ecologic Institute, Berlin.

Table 2.1 Legislative instruments of the European Union that have been identified as relevant for soil protection (taken from Frelih-Larsen et al. 2016)

Legislative tool	Cluster
Strategic initiatives	
Climate change adaptation strategy (COM / 2013/0216)	Climate change and energy
Biodiversity strategy (COM / 2011/0244 final)	Biodiversity, planning and soil sealing
Circular Economy Action Plan (COM / 2015/0614)	General tools
Seventh environmental action program (Decision No. 1386/2013 / EU)	General tools
Forestry strategy (COM / 2013/0659)	CAP and complementary measures
Roadmap towards a resource-efficient Europe (COM / 2011/0571)	General tools
Soil sealing guidelines (European Commission, 2012)	Biodiversity, planning and soil sealing
Thematic strategy for soil (COM (2006) 231)	General tools
Binding measures—Directives, regulations, decisions	
Drinking water directive (COM (2017) 0753 - C8-0019 / 2018 - 2017/0332 (COD))	Pollution, soil contamination
Effort Sharing Decision (Decision 406/2009 / EC)	CHANGING the climatic and energy
Environmental Impact Assessment Directive for Certain Public and Private Projects—EIA (2014/52 / EU)	Biodiversity, planning and soil sealing
Environmental Liability Directive (2004/35 / EC)	Soil contamination
Fertilizer Regulation (COM (2016) 157)	Pollution, soil contamination
Flood Risk Assessment and Management Directive (2007/60 / EC)	Biodiversity, planning and soil sealing
Groundwater Protection Directive from Pollution and Deterioration (2006/118 / EC)	Pollution, soil contamination
Birds Directives (79/409 / EEC) and Habitats (92/43 / EEC)	Biodiversity, planning and soil sealing
Industrial Emissions Directive (2010/75 / EU)	Pollution, soil contamination
Landfill Directive (1999/31 / EC)	Pollution, soil contamination
Regulation of greenhouse gas emissions and removals resulting from land-use activities, land-use change and forestry (529/2013 / EU)	CHANGING the climatic and energy
Mercury regulation (2017/852 / EU)	Pollution, soil contamination
Directive on national emission limits for certain air pollutants (2016/2284)	Pollution, soil contamination
Nitrates Directive (91/676 / EEC)	Pollution / CAP
Pesticides Directive (2009/128 / EC)	Pollution / CAP
Renewable energy directive (2009/28 / EC)	CHANGING the climatic and energy

(continued)

Table 2.1 (continued)

Legislative tool	Cluster
Directive on the use of sewage sludge in agriculture (86/278 / EEC)	Pollution, soil contamination
Strategic Environmental Assessment Directive—SEA (2001/42 / EC)	Biodiversity, planning and soil sealing
Waste Framework Directive (2008/98 / EC)	Pollution, soil contamination
Water framework directive (2000/60 / EC)	Pollution, soil contamination
Financing instruments	
Cohesion Fund	Pollution, soil contamination
Common Agricultural Policy (CAP)	CAP and complementary measures
European Regional Development Fund (ERDF)	Pollution, soil contamination
European Social Fund (ESF)	Pollution, soil contamination
Horizon 2020	General tools
Life + program	General tools
State aid for the environment and energy 2014–2020	Pollution, soil contamination

- Providing financial support (Life + program);
- Dedicating explicitly to soil protection (Thematic strategy for soil protection).

The Thematic Strategy for Soil Protection is the only political tool in this cluster explicitly dedicated to soil protection at the European level. The proposal for a Framework Directive accompanying this document explicitly considers soil sealing as a threat to the soil, inviting the Member States to take appropriate measures to limit its waterproofing or mitigate its effects. In addition to the thematic strategy for soil, the group includes policy instruments aimed at improving the implementation of current legislation, proposing new legislation or further integrating soil considerations. The Seventh Environmental Action Program, in particular, aims to promote the implementation of environmental legislation which can contribute to the protection of natural capital and which considers climate change as a critical factor for the soil. It promotes the integration of soil-related considerations into agriculture and forestry, in the field of renewable energy and Member States' planning. The Seventh Environmental Action Program also encourages the EU and the Member States to reflect on the introduction of a risk-based approach within a binding legal framework on soil (Mavrakis et al. 2015).

The Roadmap to a resource-efficient Europe promotes the conservation: (1) of natural capital, including the maintenance of soil fertility; (2) of the soil, in terms of reduction of erosion and loss of organic matter. Finally, other instruments such as the LIFE + Program and Horizon 2020 can contribute to soil protection through the funding of relevant research and innovation or to the action of civil society and industry. Soil results ultimately depend on the prioritization of issues related to soil in the EU agenda and the context of any initiatives presented by Member States (C. Ferrara et al. 2014b). As can be deduced from Table 2.2, the cluster considers all soil functions, although some implicitly.

The strength of this group of legislative instruments is represented by the fact that they refer to more stringent legislation or to further research relating to soil protection (Luca Salvati and Carlucci 2015). The weak point is the lack of mandatory soil requirements, mainly due to the non-binding nature of the legislative instruments considered (Francaviglia et al. 2019). This also implies that the approach and the level of ambition of soil-relevant measures and projects mainly depend on the will of the Member States to implement dedicated or specific legislation (e.g. the proposed Framework Directive on soil, annexed to the Thematic Strategy). The key strategy papers were all drafted before the withdrawal of the proposed Soil Framework Directive, therefore it is unclear how certain actions or objectives will be met and how efforts to integrate soil into EU policy will be coordinated. Furthermore, no tools within the cluster explicitly support, define or establish requirements for the protection of soil functions (Bajocco et al. 2012). Given the limitations in terms of coverage of this cluster at the EU level, it is clear that different strategies are pursued between Member States (Table 2.3). These, both for the definition of soil protection in national legislation and monitoring approaches, are conceptually different and emphasize different aspects of soil protection (Forino et al. 2015).

2.1.1.2 Cluster "Common Agricultural Policy (CAP) and Complementary Measures"

The CAP is an important economic engine for decisions affecting agriculture across the EU and has the potential to promote soil protection in both agriculture and forestry through the implementation of its measures and related obligations by Member states (Kutter et al. 2011; Kertész and Madarász 2014; M. Karamesouti et al. 2015). Although the EU level of competence for forestry is much more limited than it is for agriculture, the implementation of the EU Forestry Strategy (COM/2013/0659) is closely linked to the CAP, which remains the only source of EU funding to encourage reforestation and sustainable forest management. However, in the many Member States, national forestry policies have a more important influence than CAP on forest land management (Ferrara et al. 2017).

The CAP has three general objectives—(i) sustainable food production, (ii) sustainable management of natural resources and iii) balanced development of the territory—which are linked to those of Europe 2020 for smart, sustainable and inclusive growth (Fig. 2.3). CAP measures are the result of a series of incremental reforms since the policy was first introduced in 1962, and some soil-related measures have been available for decades. The three most relevant CAP tools for soil protection are the Good Agricultural and Environmental Condition—GAEC conditionality standards, the "greening" of First Pillar direct payments and a wide range of measures in the Rural Development Programs—RDPs (Second Pillar).

Table 2.2 The soil functions toward which the European policies of the "general instruments" cluster are addressed

Function	Thematic strategy for soil protection	Seventh environmental action program	Roadmap toward an efficient Europe in the use of resources	Circular Economy Action Plan	Life + program	Horizon 2020
Carbon pool	All soil functions have been recognized by the attached Framework Directive proposal, then withdrawn	I—Through the implementation of Member States' legislation on climate change	I—Through the implementation of Member States' legislation on climate change	N/A	I—By carrying out the activities of the Member States under the Climate Action subprogram, priority area "Mitigation of climate change"	I—Through the commitment of projects focused on soil management concerning food safety, forestry, sustainable agriculture, marine and maritime research and the bio-economy, environment, climate, efficiency of resources and raw material

(continued)

Table 2.2 (continued)

Function	Thematic strategy for soil protection	Seventh environmental action program	Roadmap toward an efficient Europe in the use of resources	Circular Economy Action Plan	Life + program	Horizon 2020
Platform for human activities	Like above	I—Through the implementation of Member States' legislation and industrial waste	N/A	THE	I—Through the commitment of the activities of the Member States in the context of the subprogram for the Environment, Thematic Priorities for Waste and the Efficiency of Resources, for the implementation of the Roadmap for an efficient Europe in terms of resources and the Seventh Environmental Action Program	Like above
Biomass production	Like above	I—Through the implementation of the legislation of the Member States on the use of raw materials, including renewable energy	I—Through the implementation of the legislation by the Member States concerning the supply of raw materials, also concerning the production of energy	By encouraging measures and legislation for the reuse of waste and raw materials, further use of soil for biomass production could be avoided	I—Through the commitment of the activities of the Member States within the subprogram for the Environment, Thematic Priorities for Resource Efficiency, the implementation of the Roadmap for a resource-efficient Europe and the Seventh environmental action program	Like above

(continued)

Table 2.2 (continued)

Function	Thematic strategy for soil protection	Seventh environmental action program	Roadmap toward an efficient Europe in the use of resources	Circular Economy Action Plan	Life + program	Horizon 2020
Biodiversity pool	Like above	I—Through the commitment of the activities of the Member States within the subprogram Environment, Priorities in Nature and Biodiversity area, for the implementation of the Birds and Habitats Directive	I—Through the integration of biodiversity and the conservation of protection values, which include soil, in other EU policies and their implementation	N/A	I—Through the commitment of the activities of the Member States within the subprogram Environment, Priorities in Nature and Biodiversity area, for the implementation of the Birds and Habitats Directive	Like above
Approval of mineral resources and raw materials	Like above	I—Through the implementation by the Member States of legislation relating to the supply of raw materials, including the Renewable Energy Directive	I—Through the implementation of legislation by the Member States on agriculture and fisheries	By encouraging measures and legislation for the reuse of waste and raw materials, further use of soil for biomass production could be avoided	I—Through the commitment of the activities of the Member States within the subprogram for the Environment, Thematic Priorities for Resource Efficiency, for the implementation of the Roadmap for a resource-efficient Europe and the Seventh environmental action program	Like above

(continued)

Table 2.2 (continued)

Function	Thematic strategy for soil protection	Seventh environmental action program	Roadmap toward an efficient Europe in the use of resources	Circular Economy Action Plan	Life + program	Horizon 2020
Storage, filtering and transfer of nutrients and water	Like above	I—Through the implementation by the Member States of the legislation relating to water and nutrients, including the Water Framework Directive, Urban Wastewater Directive, Nitrates Directive, Marine Strategy Framework Directive and Directive on Floods	E—Member States set water efficiency targets by 2020 also concerning river basin management	I—The Commission is responsible for taking a series of actions to promote the reuse of treated wastewater, including legislation on minimum requirements for water reused in agriculture	I—By carrying out the activities of the Member States within the subprogram Environment, Thematic priorities for water, the implementation of the Water Framework Directive, Floods Directive and Marine Strategy Framework Directive	Like above

I = implicitly, E = explicitly

Table 2.3 Correspondence between the issues to be explored in the policies of the "general instruments" cluster and the approach of the individual Member States

EU level issue	Summary of national approaches	Open questions
A strategic policy for soil protection	National approaches are very varied. Only a small part of the Member States have adopted strategic, binding policies for soil protection. Most of them addressed the issue of soil protection only on a partial basis, considering only a few threats or adopting a fragmented approach in integrating with policies that explicitly concern other sectors	There, it remains the question of how to represent the main objectives of soil protection in European policies and on what basis they should be integrated into other policy areas
Clear conceptualization of soil-related issues and their value	Different and evolving based on the value attributed to land and society	Need for further dialogue on how the need to protect the soil could be conceptualized. This is important for soil protection but also for the achievement of other objectives such as climate change
Coordinated monitoring of soil conditions	It derives in some ways from the conceptualization of soil issues. In the Member States, where the approach to soil protection is more clearly defined, efforts to evaluate and review soil parameters are more coordinated. Some Member States have partial systems or indicator systems under development	In the absence of a clear conceptualization of soil protection issues, there is still a long way to go in coordinating the monitoring of its condition

The main threats contemplated by this cluster are connected to soil erosion, loss of organic matter and carbon content, and indirectly to loss of biodiversity. The main soil functions toward, which this cluster is directed, are closely related to the maintenance of fertility, resilience and the productive capacity of the land for agriculture and forestry. These features are useful in addressing the main threats identified for this cluster, because many of the forest management practices that limit erosion risks and protect/improve soil carbon content also increase carbon storage capacity promote biodiversity and soil biomass production (Colantoni et al. 2015b).

The main strength of this cluster lies in the scope of its application (potentially the CAP applies to all EU agricultural and forestry land) and in the economic support to the policies for the management of the territory by farmers and owners of the forests. This applies in particular to the GAEC rules and the "greening" requirements of the First Pillar and more locally to the management of the territory and the investment

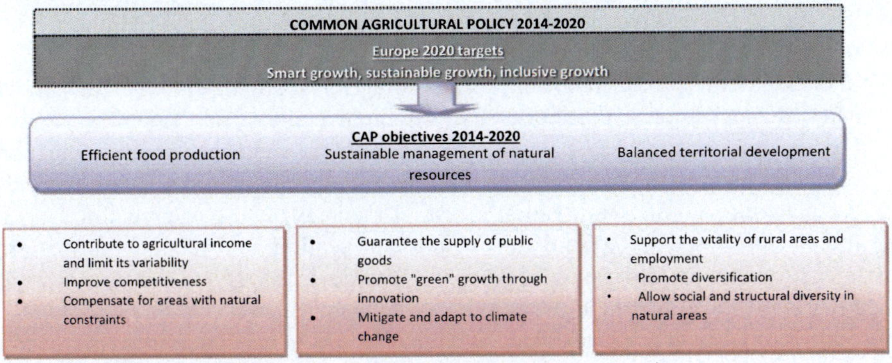

Fig. 2.1 Objectives of the Common Agricultural Policy

measures of the RDPs. The EU forestry strategy strengthens the link with sustainable forest management (including soil protection) and guides the Member States in using the forestry measures of the RDPs to achieve this. A weak point is that there is no legal obligation for Member States to demonstrate that their GAEC standards and the "greening" requirements of the first pillar are targeted at specific soil needs and priorities or other environmental objectives. There are opportunities to strengthen elements of EU legislation, in particular in relation to the GAEC soil standards, which could be further clarified, and to the first pillar greening measures which could, if strengthened, contribute more to soil protection (Kelly et al. 2015) (Fig. 2.1).

The European Parliament has already started to work on the CAP reform after 2020. An own-initiative report on the Commission communication entitled "The future of food and agriculture" (Dorfmann report) was adopted on 30 May 2018 before the presentation of legislative proposals in June 2018.

2.1.1.3 Cluster "Biodiversity, Land and Soil Sealing"

This cluster considers EU tools that aim to prevent, limit, mitigate or compensate for pressures on the natural environment and their impact on the soil. These tools belong to three categories:

- regulatory and non-regulatory instruments, focused on ensuring the protection of nature in terms of species and habitats (Habitats Directive, Birds Directive and Biodiversity Strategy);
- regulatory tools, focused on determining whether projects or plans/programs have environmental implications on the soil (Directive on environmental impact assessment and Directive on strategic environmental assessment);
- regulatory tools, focused on providing an action framework aimed at flood risk management (Directive on the assessment and management of flood risks);

The most relevant threats to soil considered by the policies of this cluster concern:

- The loss of soil biodiversity, directly linked to the provisions of the Habitats Directive and the Birds Directive, indirectly to the Biodiversity Strategy. Furthermore, in the Guidelines for the control of soil sealing it is recognized that urban sprawl and soil waterproofing constitute threats to biodiversity.
- Floods and landslides. The Flood Risk Assessment and Management Directive explicitly sets out the requirements that aim to reduce flood risks. Effective land-use planning and flood mitigation measures (such as the use of green infrastructure) are proposed in the Guidelines against soil sealing. Soil-related information, as required by the Environmental Impact Assessment Directive and the Strategic Environmental Assessment Directive, can help address this soil threat.
- Soil sealing. Explicitly considered as a threat of soil in the Guidelines against soil sealing, potentially mitigated through the use of green infrastructures, as established in the European Flood Directive, and potentially subject to an environmental impact assessment or a strategic environmental assessment.

The cluster covers all soil functions (Table 2.4), giving particular importance to the conservation of biodiversity and support for human activities.

This set of policies offers opportunities to improve actions aimed at protecting soil biodiversity and against soil sealing (Table 2.5). However, at the moment, the legislative instruments dedicated to these purposes are potentially limited by the factors listed below:

- A strategic vision for soil protection is missing that adequately takes into account its functions and that lays the foundations for integrating soil-related actions and soil biodiversity into broader policies.
- Soil biodiversity is only implicitly, not explicitly covered by binding measures for nature conservation. Furthermore, indications of the possibilities of action in favor of soil biodiversity are scattered among the various legislative documents.
- Soil sealing is considered explicitly only by non-binding guidelines and is partially related to the flood directive. However, there are areas where soil waterproofing can take place through other political priorities, from the development of infrastructure to remediation practices. A better understanding of the integration of soil sealing problems at EU level would be useful as part of a broader message on soil protection.

The problem of soil sealing is closely linked to the internal settlement pressures of each Member State and to the planning of their territory, which remains a national/regional competence, so that at EU level the possibilities for intervention, at the moment, have been more limited.

2.1.1.4 Cluster "Climate Change, Energy"

This cluster focuses on policies dedicated to climate change and the contribution that soil and land management can make to mitigate climate change and support climate adaptation activities. A key function of the soil is the conservation and release of

Table 2.4 The soil functions toward which the European policies of the "biodiversity, planning and soil sealing" cluster are addressed

Function	Habitat Directive	Birds Directive	Biodiversity strategy	Soil sealing guidelines	Floods Directive	EIA Directive	SEA Directive
Carbon pool	Protect and restore carbon-rich habitats, such as peat bogs and other wetlands	Protect and restore carbon-rich habitats, such as peat bogs and other wetlands	I—The implementation by the Member States of support actions and conservation measures under Objectives 1, 2, 3 and 6 can contribute to the protection of carbon in the soil	E—The guidelines recognize that soil sealing affects carbon sequestration and storage	N/A	I—The EIA requires a description of the factors that can be significantly influenced by projects potentially impacting on the soil as a carbon reserve	I—The SEA requires a description of the possible impacts on the environment potentially connected to soil protection as a carbon reserve

(continued)

Table 2.4 (continued)

Function	Habitat Directive	Birds Directive	Biodiversity strategy	Soil sealing guidelines	Floods Directive	EIA Directive	SEA Directive
Platform for human activities	Some urban green areas are also Natura 2000 sites but are not adequately protected by the Directive	Like above	I—Member States' implementation of support actions and conservation measures under Objectives 1, 2, 3 and 6 can help protect the soil as a platform for human activities	E—The guidelines suggest that green infrastructures can mitigate the effects of soil sealing and therefore support this soil function	N/A	I—The EIA requires a description of the factors that can be significantly influenced by projects potentially impacting on the soil as a platform for human activities	I—The SEA requires a description of the possible impacts on the environment that are potentially linked to the soil as a platform for human activities
Biomass production	Many forests (which produce biomass) are protected by Natura 2000	Like above	I—The implementation by the Member States of support actions and conservation measures under Objectives 1, 2, 3 and 6 can contribute to the production of biomass from the soil	N/A	N/A	I—The EIA requires a description of the factors that may be significant—the mind influenced by potentially impacting projects on the ground for the production of biomass	I—The VAS requires a description of the possible impacts on the environment potentially—mind connected to the protection of the soil for the production of biomass

(continued)

Table 2.4 (continued)

Function	Habitat Directive	Birds Directive	Biodiversity strategy	Soil sealing guidelines	Floods Directive	EIA Directive	SEA Directive
Biodiversity pool	I—The designation by Member States of Natura 2000 sites and the implementation of the related conservation measures can favor this function	I—The designation by Member States of Special Protection Zones and the implementation of the relative conservation measures can favor this function	I—The implementation by the Member States of support actions and conservation measures under Objectives 1, 2, 3 and 6 can contribute to the conservation of biodiversity	E—The guidelines recognize that soil sealing affects biodiversity both in the soil and above the soil	N/A	I—The EIA requires a description of the factors that may be significant—the mind influenced by potentially impacting projects on the ground as a reserve for biodiversity	I—The VAS requires a description of the possible impacts on the environment potentially—mind linked to soil protection as a reserve for biodiversity
Approval of mineral resources and raw materials	Some Natura 2000 sites allow the extraction of raw materials, provided that the species and habitats are not adversely affected	Like above	I—The implementation by the Member States to support actions and conservation measures under Objectives 1, 2, 3 and 6 can contribute from the ground	E—The guidelines recognize that the soil sealing affects many compromised fertile areas—tend the soil in terms of safety	N/A	I—The EIA requires a description of the factors that may be significant—the mind influenced by potentially impacting projects on the ground as a reserve of raw materials	I—The VAS requires a description of the possible impacts on the environment potentially—mind linked to soil protection as a reserve of raw materials

(continued)

Table 2.4 (continued)

Function	Habitat Directive	Birds Directive	Biodiversity strategy	Soil sealing guidelines	Floods Directive	EIA Directive	SEA Directive
Preservation, filtering and transfer of nutrients and water	N/A	N/A	I—Member States' implementation of support actions and conservation measures under Objectives 1, 2, 3 and 6 can contribute to soil conservation and filtering of water and nutrients	E—Through the Guidelines, it is recognized that the removal of the upper soil layer can hinder the filtering of rainwater and modify ecosystems and services connected to water	I—Planning rules can allow better soil management and the construction of green infrastructure to control water runoff	I—The EIA requires a description of the factors that may be significant—mind influenced by potentially impacting projects on the soil and its ability to retain and filter water	I—The VAS requires a description of the possible impacts on the environment potentially connected to the protection of soil and of its ability to retain and filter the water

I = implicitly, E = explicitly

Table 2.5 Strengths, weaknesses, opportunities and threats in relation to soil protection: the cluster of policies on "biodiversity, planning and soil sealing"

Legislative instrument	Strengths	Weaknesses	Opportunity	Threats
Habitat Directive	The Habitats Directive establishes an action framework for the protection of biodiversity across the EU, which can indirectly contribute to addressing a range of soil threats (loss of organic matter, contamination, erosion, compaction and loss of biodiversity) through protection and restoration of semi-natural and natural habitats	The directive does not explicitly establish mandatory soil requirements	Member States can choose from a wide range of actions, including the reduction of intensive agriculture and habitat fragmentation, which have several beneficial effects on the soil	The threats do not reside in the Directive but it does not implement relevant conservation measures for the protection of soil

(continued)

Table 2.5 (continued)

Legislative instrument	Strengths	Weaknesses	Opportunity	Threats
Birds Directive	The Birds Directive establishes a framework for action for the conservation of all wild species of birds in the wild in the EU, which can indirectly contribute to addressing a range of soil threats (loss of organic matter, contamination, erosion, compaction and loss of biodiversity) through the protection and restoration of avifauna habitats	The directive does not explicitly establish mandatory soil requirements	When defining conservation measures for each SPA, Member States may choose between the measures deemed most appropriate, including those which have potentially beneficial effects on soil	The threats lie not in the directive itself but the non -implementation of conservation measures relevant for soil protection
Biodiversity strategy	The Strategy sets a long-term vision by 2050 and a goal to be achieved by 2020 for the conservation of biodiversity within the EU, which have positive implications for a broad spectrum of threats and soil functions	The Directive does not contain mandatory or optional objectives explicitly focused on the soil	The intervention scale of the network Natura 2000 conservation measures, and integrated more issues related to biodiversity and services ecosystem in agriculture and Forestry	The threats do not reside in the same strategy by the Member States
Soil sealing guidelines	The guidelines provide a detailed set of good practices and examples to limit, mitigate and compensate the effects of soil sealing in the EU, highlighting that it is possible to implement the principles for sustainable land use only through correct and careful territorial, regional planning and local	The guidelines do not contain mandatory requirements	In implementing measures to limit, mitigate or compensate for soil sealing, Member States have some degree of flexibility in the implementation of certain types of measures and the appropriate level of governance	The threats do not reside in the same guidelines, but not in the implementation of good practices

(continued)

Table 2.5 (continued)

Legislative instrument	Strengths	Weaknesses	Opportunity	Threats
Flood Risk Directive	The Directive establishes an EU-level approach to flood risk management, with benefits for soil protection if implemented	The directive does not establish mandatory soil-focused requirements	Member States can put in place additional voluntary or mandatory actions to be implemented in the Directive, aimed at supporting soil protection	The threats lie not in the policy itself, but in the failure of Member States to implement measures to protect the soil
EIA Directive	The EIA establishes a framework for action to determine whether a project needs to undergo an evaluation, which includes information on possible impacts on soil and alternative practices before authorization for its development is granted	The directive does not explicitly set mandatory soil targets or mandatory requirements	When defining alternatives if a project affects soil quality, more suitable measures may be chosen in addition to the requirements set out in the directive, aimed at ensuring a higher level of soil protection	The threats do not reside in the Directive but it does not implementation of the requirements by the Member States
SEA Directive	The SEA establishes a legislative framework to assess the environmental effects of certain plans and programs undertaken by the Member States. The related report must contain information on significant possible effects, including on soil	It does not explicitly establish binding soil requirements. There is no mechanism established by the directive to prevent further soil degradation due to certain plans or programs	Member States can choose the most appropriate actions for soil protection in response to the possible environmental impacts of implementing a plan or program	The threats lie not in the directive itself but the non-implementation of the requirements by the Member States

organic matter and carbon. Soils can compensate for greenhouse gas emissions by capturing and storing carbon (albeit reversibly) and can contribute to adaptation to climate change, for example through flood mitigation (Luca Salvati et al. 2008; Henderson et al. 2015; Green et al. 2017). Mitigation and adaptation to climate change are certainly influenced by threats to the soil, including the loss of organic carbon, which leads directly to the emission of carbon into the atmosphere (Novara et al. 2013; Müller-Nedebock and Chaplot 2015; Colantoni et al. 2016). There is also a connection between soil organic matter with other threats such as compaction, erosion and flooding (Cecchini et al. 2019).

This cluster includes four EU legislative instruments, which do not have soil management and protection as their ultimate goal, even if the management practices they promote contribute to its achievement. These policies are:

- The Effort Sharing Decision (ESD). It sets binding national targets for greenhouse gases, for each of the Member States of the European Union, which total a 30% reduction in emissions by 2030 (compared to a 2005 reference base). This tool is linked to soil protection given the connection between soil management and greenhouse gas emissions associated with agriculture.
- Soil and forestry regulation for 2021–2030. Under EU legislation adopted in May 2018, EU Member States must ensure that greenhouse gas emissions from land use, land-use changes or forestry are offset by at least one removal equivalent of CO_2 from the atmosphere in the period from 2021 to 2030. The field of application is extended not only to forests but to all land uses.
- The strategy of adaptation to climate change. It provides a general framework to increase adaptation through voluntary mechanisms that improve preparation and the ability to respond, at different levels, to the effects of climate change.
- The Renewable Energy Directive. The spread of the use of renewable energies, if it involves the consumption of soil or a change in land management, has a potential impact on soil protection. It is most relevant for the soil in the context of the expansion of bioenergy and the use of biofuels.

The policies analyzed in this cluster have been linked to soil carbon and biomass production functions (Table 2.6). The only soil function that is explicitly linked to these policies is the maintenance of the carbon pool in the soil. This link is contained in the Renewable Energy Directive, as a result of clauses that prevent the exploitation of areas with a high carbon content for the production of biofuels.

One of the strengths of the policies examined in the context of this cluster is the opportunity they offer in promoting holistic soil health. The question of the organic matter content, linked to the objectives of climate mitigation and adaptation, offers great benefits for the health of the soil, in terms of improving its structure and reducing the susceptibility to compaction and erosion. However, the emphasis is on reducing emissions rather than holistic soil management. There is a risk that soil protection may fall into the background in the absence of clear objectives or priorities dedicated to this resource (Table 2.7).

Table 2.6 The soil functions toward which the European policies of the "climate change, energy" cluster are directed

Function	Adaptation strategy	Renewable energy directive	Effort sharing decision	Soil and forestry regulation
Carbon pool	Implicit link to the presence in the LIFE program of action priorities toward adaptation to climate change	The Directive is explicitly aimed at reducing greenhouse gas emissions. Furthermore, there are specific provisions aimed at limiting the consequences of the use of biofuels at least for areas with high carbon content	Implicit link	E—Through the maintenance and storage of carbon
Platform for human activities	I—Through the implementation of the Rural Development Programs	N/A	N/A	N/A
Biomass production	I—Through the implementation of the Rural Development Programs and by integrating adaptation measures into the EU forestry policy and the legislation of the Member States	I—The link between renewable energy and biomass production is evident	I—The adoption of agricultural practices that limit greenhouse gas emissions take part in the achievement of more general objectives within the ESD	I—The maintenance of the carbon pool potentially impacts on how biomass is produced and, in the long term, also on the state of health of the soils for biomass production
Biodiversity pool	I—Through the integration of adaptation measures into EU biodiversity policy and Member States' legislation	N/A	N/A	N/A
Approval of mineral resources and raw materials	Implicit link through the implementation by the Member States of rural development programs under the second pillar of the CAP	I—The link between renewable energy and biomass production is evident	N/A	N/A

(continued)

Table 2.6 (continued)

Function	Adaptation strategy	Renewable energy directive	Effort sharing decision	Soil and forestry regulation
Storage, filtering and transfer of nutrients and water	I—Support for Member States' actions concerning flood risk planning and management	N/A	Implicit link regarding changes in nutrient management practices	I—The improvement of organic matter should improve the structure of the soil and in particular the filtering of water and nutrients

I = implicitly, E = explicitly

Table 2.7 Strengths, weaknesses, opportunities and threats in relation to soil protection: the cluster of policies on climate change and energy

Legislative instrument	Strengths	Weaknesses	Opportunity	Threats
Climate change adaptation strategy	The strategy provides a general framework for increasing adaptation through voluntary mechanisms aimed at improving preparedness and the ability to respond, at different levels, to the effects of climate change, which also affect the soil	There are no specific mandatory requirements on the soil	Member states are provided with a wide range of voluntary tools (guidelines, reports, monitoring, funding, development of indicators) to increase national adaptation strategies, which can have positive impacts, direct or indirect, on soil protection	They lie not in the adaptation strategy itself, but by the inaction of the Member States
Renewable energy directive	Offers some protection of particularly vulnerable and carbon-rich soils. However, this protection is only offered in terms of expanding the storage of raw materials for biofuels	No specific requirements are governing the use or protection of the soil if not strictly connected to the areas from which liquid biofuels are to be obtained	The criteria for the use of degraded land can be improved and updated	They derive from the extensive use of biomass to derive energy
Effort Sharing Decision	It forms a basis for addressing issues related to greenhouse gas emissions concerning soil management	The current decision is very open in terms of the actions a member state can take to reduce greenhouse gases. Therefore, actions to address soil management and aspects relevant to agricultural emissions vary between Member States	The tightening of targets for 2030 offers the opportunity to focus more on reducing emissions	Many alternative strategies can be used to reduce emissions. Without a policy that guides the importance of action on the ground, it is unlikely that this can be coordinated in the future
Soil and forestry regulation	It allows improving soil monitoring, defines/promotes the adoption of specific "measures" and management practices	N/A	It offers the opportunity to consider soil management more holistically	N/A

2.1.2 European Urban Agenda: Sustainable Land Use and Nature-Based Solutions

The European Union's agenda for research and innovation policy on nature-based solutions and the renaturalization of cities aims to position the EU as a leader in the process of "innovation with nature", for more sustainable societies and resilient. The concept of "Nature-Based Solutions" (NBS) is relatively new and has been introduced to promote nature as a means of providing solutions in mitigation measures and adaptation challenges to climate change (Cohen-Shacham et al. 2016). It is now part of the new framework program for research and innovation, Horizon 2020, linking the concepts of biodiversity and ecosystem services to the innovation objectives for (Maes and Jacobs 2017) economic growth and job creation, and opening paths of transformation toward sustainable social development (Maes and Jacobs 2017).

In essence, solutions to social challenges based on nature can be defined as solutions that are inspired by and supported by nature, economically advantageous, capable of simultaneously offering environmental, social and economic benefits, helping to create resilience (Fini et al. 2017; Keesstra et al. 2018; Pili et al. 2019). These solutions bring nature and natural elements into cities and metropolitan landscapes through interventions adapted to the local context, resource-efficient and systemic (Biasi et al. 2015). This purpose derives from the growing recognition and awareness that nature can help to provide valid solutions capable of using and intelligently "engineered" the services and benefits of natural ecosystems, providing sustainable, convenient, multifunctional and suitable alternatives for various purposes. Working with nature, rather than against it, can facilitate the efficiency of the economy in terms of resources, make it competitive and greener. Among the specific objectives of the NBS, there is the restoration of degraded ecosystems and the improvement of the sustainability of urbanization (Kalantari et al. 2018). In this context, particular attention is given to the phenomenon of urban sprawl, the redevelopment of abandoned areas and the renaturalization of urban areas in general (Carlucci et al. 2018; Ciommi et al. 2018; Luca Salvati et al. 2018a).

More than half of the global population now lives in urban areas and it is known that this figure will increase; it, therefore, becomes a priority to think about solutions that can allow more sustainable urbanization (Luca Salvati et al. 2017a). The same phenomenon of urban sprawl, which consumes large quantities of soil and resources in a dispersive and undifferentiated way, represents a challenge in this area (Oueslati et al. 2015). All of Europe needs forward-looking and environmentally friendly approaches in favor of the soil and its management, including at the political level. The research, through the NBS, offers the possibility of providing innovative and effective methods for working in this direction, both in terms of knowledge and sharing of good practices and dissemination of tools and technologies to local actors (useful for example to those who deals with planning). First, they support economic development in urban areas that depend on the quantity and quality of the natural resources available and which can guide new business models, which decouple economic growth from the depletion of resources and their unfair distribution (Ciommi et al. 2017a). This can favor the circular economy and increase dependence on local resources, leading to greater efficiency in the consumption of energy and materials (Ciommi et al. 2017b).

Secondly, planning that takes NBS into account has a positive environmental impact, for example in terms of mitigation toward climate change, increasing urban resilience to environmental risks, such as drought, floods and heatwaves. It can also reduce pressure on peripheral natural areas, for example by treating wastewater at a shorter distance from sources (Oral et al. 2020). NBS has an important role, for example, in supporting the implementation and optimization of green, blue and grey infrastructures. The enhancement of green infrastructures can contribute to reducing the needs and costs of energy and resources (circular economy) thanks to the cooling and thermal insulation provided by the plant system, which also reduces the "urban heat island" effect or green roofs and vertical green can also reduce the need for heating and air conditioning (Pearlmutter et al. 2020). In line with the NBS, a new attitude emerges from the planners of the territory who are trying to exploit the space more effectively through the search for new uses for the soil and for the grey infrastructure that are underutilized or unused. The possibilities for sustainable urban growth can be found, for example, in the conversion of abandoned land into urban farms and community gardens, and in the regeneration of former industrial sites, through the bioremediation of toxic soils and their subsequent transformation into green spaces.

To conclude, EU policy places European cities as laboratories for innovation, experimentation and to test the effectiveness of nature-based solutions, maximizing a variety of environmental, social and economic benefits (Nesshover et al. 2017). Existing urban networks can facilitate the dissemination of demonstration projects and the capacity for intervention on a larger scale.

2.1.3 United Nations Guidelines for Sustainable Land Use

Among the initiatives through which the United Nations intend to promote sustainable land use are (1) the Human Settlements program and (2) the sustainable development objective no. 15 and in particular the target 15.3.

The purpose of the United Nations Human Settlements (UN-HABITAT) program is to promote sustainable urbanization both from a social and environmental point of view and to guarantee everyone the right to have a dignified home. UN-HABITAT is the main agency that within the United Nations system coordinates activities in the area of urban agglomerations and other human settlements.[10] In this historic moment in which we are witnessing unprecedented growth in cities, sustainable urbanization represents one of the most urgent challenges for the global community of

[10] The organization was founded in 1978 because of the first conference on human settlements and sustainable urban development (Habitat I) in Vancouver, Canada in 1976.

the twenty-first century. In its activity aimed at improving human settlements in terms of sustainability and respect for the environment, reducing poverty and increasing safety in the city, it was essential for UN-HABITAT to cooperate with many actors, forging relationships with the civil society, various municipalities and regional and international authorities (Ciommi et al. 2017b). This agency was also assisted by non-governmental organizations, urban professionals and researchers, developing profitable agreements with the governments and municipalities they belong to at a regional level, the cooperation concerns the coordination and enhancement of the communication flow, the exchange of information, data sharing and the facilitation of the monitoring and evaluation of the projects and programs being implemented. Regional cooperation also promotes the creation of national and regional networks of civil society organizations and other partners. At the international level, cooperation focuses on policy formulation, resource sharing, promotion and awareness-raising activities (United Nations 2011).

Sustainable use of land is also an integral part of the 2030 Agenda, or the "action program for people, the planet and prosperity" (from the Preamble to the 2030 Global Agenda). Signed in September 2015 by the governments of the 193 member countries of the UN, this program encompasses 17 Sustainable Development Goals—SDGs—for a total of 169 targets, offering new opportunities for an ambitious and integrated environmental policy worldwide and in the European Union.

Among the desired goals that directly or indirectly affect the soil, governments should, by 2030, improve the sustainability of the current urban development model and planning tools[11] (i) ensure universal access to green spaces and public spaces safe, inclusive and accessible[12] (ii) Nevertheless, the most relevant for soil is target 15.3 which aims to achieve "a land degradation-neutral world" by 2030 (Stavi and Lal 2015). Many other targets of the SDGs are however relevant for achieving sustainable management of terrestrial systems, emphasizing the importance of ecological restoration in degraded production landscapes. The same target 15.3 participates at the same time in the achievement of other SDGs, including those related to mitigation and adaptation to climate change, the conservation of biodiversity, the restoration of ecosystems, the safety of food and water, the reduction of the risk of disasters and poverty.

The LAND DEGRADATION NEUTRALITY (LDN) represents a paradigm shift in territorial management policies and practices, being a unique approach that aims to offset the estimated loss of productive land through the recovery of degraded areas (Akhtar-Schuster et al. 2017). In particular, objective 15.3 aims to "guarantee the fight against desertification, the restoration of degraded land and soil, including those affected by desertification, drought and floods". According to the definition provided by the United Nations Convention to Combat Desertification (UNCCD) and approved during the 12th Conference of the Parties (UNCCD 2015), the LDN

[11] 11.3—"*By 2030, enhance inclusive and sustainable urbanization and capacity for participatory, integrated and sustainable human settlement planning and management in all countries*".

[12] 11.7—"*By 2030, provide universal access to safe, inclusive and accessible, green and public spaces, in particular for women and children, older persons and persons with disabilities*".

is a "state in which the quantity and quality of the soil resource necessary to support the functions and services of ecosystems and to improve food security, they remain stable or increase within a specific time or space scales". The use of the words "quantity" and "quality" suggests that both qualitative and quantitative measures and indicators are useful in assessing progress toward LDN. This definition also uses the words "ecosystem functions and services", which are common within the Convention on Biological Diversity and the United Nations Framework Convention on Climate Change (Costanza et al. 1998; Gómez-Baggethun et al. 2010), but rather complex for those who manage the territory by operating at the level of actions aimed at preventing, reducing and reversing land degradation. Despite this operational difficulty, target 15.3 has become a valuable vehicle for implementing the implementation of the UNCCD (2015): at the twelfth session of the UNCCD Conference of the Parties held in October 2015 in Ankara, the participating countries reached an agreement decisive to approve this vision and link the implementation of the Convention to target 15.3 of the Sustainable Development Goals (USA) Dodds et al. (2016).

2.1.4 Institutional Tools to Combat Land Degradation in Southern Europe

The identification of the areas affected by the vulnerability to soil degradation is an essential prerequisite for the fight against land degradation as it allows you to plan interventions, establish their priorities and verify their effectiveness. The UNCCD (United Nations Convention to Combat Desertification) identifies in the National Action Program (NAP) the tools through which to adopt and implement actions to combat desertification, the most evident manifestation of land degradation. They were developed through a participatory approach that involved the relevant government offices, scientific institutions and local communities. The NAPs are also linked and contain synergies with other global conventions on climate, biodiversity and water protection.

As indicated in Annex IV of the Convention against desertification, Italy and the other northern Mediterranean countries constitute within the UNCCD a regional group which aims to identify and implement common policies to combat desertification in the context of the policies of the European Union. Through the NAPs, the application and enhancement of existing national standards and intervention instruments of the European Union are proposed in these countries, encouraging the implementation of increasingly targeted laws and programs at the regional level. In addition to identifying areas vulnerable to land degradation, they contain considerations regarding land-use planning, international cooperation, proposals to restore degraded areas, the role of research, legal and institutional measures and the selection of pilot areas where apply the proposed strategies against desertification.

The action measures proposed in the Action Programs of Annex IV can be classified into four main areas of application: the agricultural sector (eg soil erosion control, salinization control), the forestry sector (e.g. protection against change land use, fire control, protection from intensive grazing, forest management), the water resources sector (e.g. measures for irrigation, for urban and industrial use and to increase the availability of water) and the social sector -economic (e.g. involvement of local populations). In particular, in the NAP adopted by Greece, these measures are divided according to the reversibility of land degradation. It is interesting to note that for the management of non-reversibly degraded areas, the application of non-rural land uses, such as urban sprawl and industrial or tourism development, is proposed for this country, while among the recovery measures in reversibly areas degraded are those of protection and are inspired by principles of sustainability (Costas Kosmas et al. 2015).

Peculiar to the action program adopted by Italy is the proposal for "rebalancing the territory": the need to intervene through adequate social and economic policies for the "productive, landscape and naturalistic recovery of areas currently compromised by an excessive concentration of activities is noted anthropic or by the abandonment of marginal areas or historic centers". In order to achieve this aim, the traditional knowledge of local populations is given great importance, which "constitutes a precious and irreplaceable patrimony of knowledge capable of suggesting how to use the potential of nature, without exhausting them". Among the measures, the following are proposed: (i) the recovery of traditional technologies and historical centers with demonstrative and proactive function of a balanced, creative and thrifty use of resources; (ii) the incentive for the adoption of urban plans that provide for the use of technologies oriented to the appropriate use of resources; (iii) the renaturalization and environmental transformation of areas subject to degradation in urban and industrial areas; and, (iv) incentives for production activities in marginal hilly and mountainous areas (Salvati et al. 2007).

The problem of land abandonment, the poor valorization of agricultural areas and the lack of protection of the territory is given particular importance in the Portuguese PAN, suggesting to promote rural and local development in the regions most vulnerable to desertification and drought. It also highlights the need to reduce anthropogenic pressure in most densely populated areas. The strategic objective shared with the NAP of Spain is to integrate the fight against land degradation into economic and social development policies, for which greater coordination is proposed.

In the action program of the latter country, where the critical nature of the risk to desertification is also well illustrated in cartographic terms (Fig. 2.2), the practices of conservation of productive soils and maintenance of those present in cultivated areas are encouraged mountain areas, also for the protection of the social and cultural value associated with them (Fig. 2.3). These practices include techniques for restoring agricultural soils and improving irrigation techniques to protect the soil from erosion.

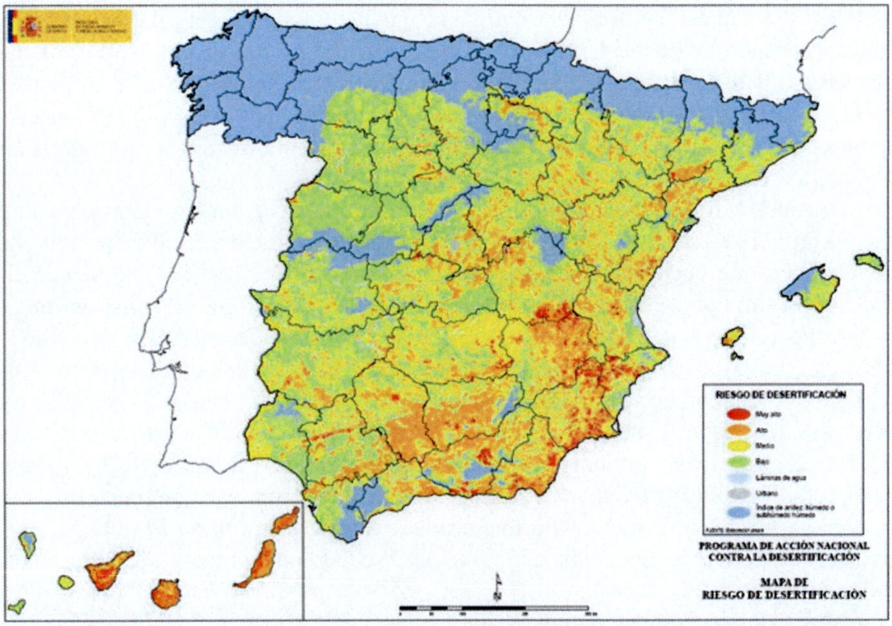

Fig. 2.2 Image is taken from the National Action Program adopted by Spain, a country where the issue of risk to desertification is quite critical

Fig. 2.3 Examples of land degradation processes in Spain located in the Sierra of Albarracín (Aragón) and Arroyo Totalán (Andalucía). Photo by Jesús Rodrigo-Comino

2.2 Quality, Biophysical Degradation, Soil Scaling: Research and Experiences

The structure and composition of the landscape continuously evolve in space and time, influencing the physical, chemical and biological processes of the soil (Canfora et al. 2017). These influences contribute significantly to the complex interactions between the natural environment and anthropic activities, shaping the characteristics and properties of the lands in various ways (the spatial diversification of the properties of the soil and its compaction are typical examples). Soil degradation processes represent a threat to the sustainability and production capacity of agricultural activity; human is among the agents that most influence it, both by strengthening and depressing it (Rodrigo-Comino et al. 2015).

In a context in which the economy, society and the environment tend to become increasingly integrated and interconnected dimensions, issues related to sustainable development are becoming increasingly important (Salvati 2014a). The debate on the organization of the territory, together with that on the design and construction of new infrastructures, is linked to these concepts, with particular emphasis in contexts characterized by an evident dispersion of the building, typically combined with fragmentation[13] phenomena of the landscape, especially of the agricultural one (Salvati et al. 2018). Urban dispersion increasingly produces a misalignment of materials and processes between urbanity and rurality and transforms the countryside into new suburbs and uncultivated spaces, depleting their resources and compromising the possibility of the inhabitants who live near these spaces and that of future generations of taking advantage of the economic and social benefits that the environment could offer them (Fig. 2.4).

Considering also these premises, preserving soils with good quality should be recognized among the main targets within the environmental protection strategies, but there is still a long way to go. At the end of this book, we would like to deepen the strategies aimed at preserving and enhancing soils with good land quality (therefore fertile), not only by proposing to integrate the principles in favor of sustainable urbanization more into the governance of the territory but also by first exploring the understanding of the processes that influence the quality and availability of productive soil.

2.2.1 *What Is Land Quality?*

LAND QUALITY is a multi-dimensional concept that represents the soil's ability to perform specific functions, continuing to guarantee the productivity of natural and/or agricultural systems.

[13] FRAGMENTATION is basically a dynamic process, of anthropic origin, which consists in the division of the natural environment into increasingly smaller and isolated fragments separated by a land cover matrix transformed by man (Forman et al. 1995; Bogaert et al. 2011). The fragmentation effect produced in an environmental mosaic varies in relation to the characteristics of the matrix. The destruction and transformation of natural environments, their reduction and increased insulation are the main components of the fragmentation process. They influence the structure and dynamics of certain sensitive animal and plant populations, to the point of altering community parameters, ecosystem functions and ecological processes. Biodiversity is reflected in the landscape: when it becomes more uniform, crops are simplified, hedges are destroyed, channeled streams, living conditions become increasingly hostile for an increasing number of species and in parallel fundamental ecological functions are lost.

Fig. 2.4 Cover of the Time magazine, published in August 2013, dedicated to a phenomenon that seems to sordidly affect our planet, the depopulation of the hives (the mysterious hive depopulation syndrome began in 2006 when, for reasons not yet fully understood, the death of bees led to the loss of ten million hives in just the United States of America. *A world without bees* would lead to and hungrier world because bees, for thousands of years, are the invisible workers who, thanks to pollination, support the agro-food system of human civilization)

A healthy urban area should at least:

- filter the water;
- maintain its shape/structure with changes in temperature and humidity;
- support the weight of the buildings.

Since constructions are often widespread in urban areas, it is important to protect the soil free of constructions, which is fundamental in providing ecosystem services and in supporting animal and plant biodiversity. Some authors (for example, Warkentin 1995) simply link the land quality to the quantity of the crop that is produced in a given plot of land. However, others emphasize the importance of how land quality affects the quality of food that derives from productive landscapes or how the environment is available to the animal and plant organisms that inhabit it (Zambon et al. 2017a).

Numerous other aspects associated with the living and dynamic nature of the soil are found if the concept of land quality is considered about the different uses of the soil: forest and pasture ecosystems, urban and industrial areas, recreational uses, etc. This implies that due to the diversity of potential land uses, the concept of "quality" should be regarded as relative rather than absolute. The suitability of soils for specific uses depends above all on intrinsic qualities, on permanent characteristics that are difficult to modify: for example, a deep soil has more space for the root system of plants than a more rocky one near the surface (Renzi et al. 2017).

Over the years the concept has therefore evolved, thanks also to authors such as Carter et al. (1997) and Karlen et al. (2003) who consider land quality as the ability of a specific type of soil to perform functions supporting the productivity of plants and animals, maintaining or increasing the quality of air and soil, supporting health and human settlements. This conception emphasizes the value that the soil has in supporting the functions of ecosystems and implies an explicit judgment on which soil conditions adapt to the principles of sustainability (Fig. 2.5). Its link with sustainability configures land quality, not as an abstract concept, on the contrary, it reinforces its management meaning (Bouma 2002). In this sense, the soil is seen as a living system that projects all its functions outward, interacting with the landscape and the environment and absorbing its transformations (Esposito et al. 2016).

To conclude, we can consider land quality as a concept that integrates the intrinsic and dynamic qualities of the soil (De la Rosa 2005) and in this thesis, in particular, it is considered at the antipodes of *land degradation*.

Fig. 2.5 Representation of a productive landscape, typical of the Mediterranean area: Vincent Van Gogh's "Olive Grove", June 1889 (oil on canvas, National Gallery of Art, Washington)

2.2.2 *Characteristics and Main Causes of Land Degradation*

Land degradation involves, among the most evident effects, a progressive reduction of soil fertility and adversely affects food safety and quality of life (Abu Hammad and Tumeizi 2012). The term land degradation has long been the subject of scientific and political debate, for example in conjunction with the themes of desertification, deforestation, soil erosion, or with certain management approaches such as "sustainable land management". With the discussion for the achievement by 2030 of "a land degradation-neutral world" in the context of the SDGs, the term land degradation has acquired a new and stronger political weight (Chasek et al. 2015). UNCCD has played an important role in defining and characterizing the term "land degradation" internationally, concerning article 1 of the UNCCD text.[14] A slightly refined definition of this definition was also included in the official definition for the implementation of objective 15.3:

> [...] Land degradation means the decrease or disappearance of biological or economic productivity and the complexity of non-irrigated cultivated land, irrigated cultivated land, pastures, forests or wooded areas following the use of lands or because of one or more phenomena, due to the activity of man and his ways of settlement.[15]

At the center of this definition, we can find the functions and the economic value of the lands for agriculture and forestry, which are promoted by human activities. Anthropogenic activities which, at the same time, can manifest themselves negatively in the form of measurable phenomena such as erosion, loss of soil productivity and loss of vegetation (Salvati and Zitti 2012; Biasi et al. 2015; Duvernoy et al. 2018).

Land degradation, which sometimes occurs without a clear understanding of the processes involved (Le Houérou 1993; Puigdefábregas and Mendizabal 1998), in arid regions, is often associated with extreme biophysical and socio-economic phenomena that can lead to an irreversible process of environmental degradation: desertification (Montanarella 2007). This term, although reporting the root "desert", is not intended as a "generator of deserts", but as reported in the United Nations Convention for the Fight against Drought and Desertification: as a serious form and irreversible land degradation.[16] From information sources of the United Nations, it emerges that 70%

[14] "[…] the expression" land degradation "means the decrease or disappearance, in arid, semi-arid and sub-humid dry areas, of biological or economic productivity and the complexity of non-irrigated cultivated land, irrigated cultivated land, pastures, forests or wooded surfaces following the use of land or one or more phenomena, in particular phenomena due to human activity and its methods of settlement, including: (i) l soil erosion caused by wind and/or water; (ii) the deterioration of the physical, chemical and biological or economic properties of the lands; (iii) the long-term disappearance of natural vegetation […] ". Definition taken from article 1 of the UNCCD.

[15] Inter-Agency and Expert Group on Sustainable Development Goal Indicators, IAEG—SDGs 2016.

[16] The operational definition of the concepts of land degradation and desertification is often accompanied by terms widely used in the scientific literature such as "sensitivity" and "vulnerability", not always used univocally or, even, used as synonyms. The need for clarity and uniqueness of the terms is particularly felt in the international context where correct communication takes on even more importance. In this sense, the IPCC (Intergovernmental Panel on Climate Change) has warned of

of the arable lands, equal to about 30% of the total of the emerged lands, are affected by land degradation phenomena and, therefore, is at risk of desertification (Salvati et al. 2013). Considering the heavy repercussions on populations, as well as on the environment, the problem is particularly serious in developing countries present in Africa, Asia, Central and South America. However, even structurally strong regions and societies (for example those present in the United States, Australia and Europe) are variously affected by the phenomenon whose evolution, although attributable to a heap of different causes, appears unequivocally connected and strengthened by arid climatic conditions and/or drought (Latorre et al. 2001; Ferrara et al. 2014a).

As shown in Fig. 2.9, the European Environment Agency supported by the ETC-TE (European Topic Center on Terrestrial Environment) has already made available a map on sensitivity to desertification and drought in Mediterranean countries, a scale of 1: 1,000,000 (European Environment Agency 200AD).

It should be noted that an articulated and exhaustive classification of the causes of land degradation is currently not achieved despite the research experiences undertaken, both nationally and internationally (Salvati et al. 2017b). Natural causes include the climate and climate change, as well as soil deterioration processes accelerated by biophysical factors; in anthropic causes, we can consider as paradigmatic examples the change of land use, urban sprawl, landscape transformations and agriculture, which must be evaluated both for the effects of pressure it exerts on ecosystems, and for the positive role of territorial protection (Kosmas et al. 2000). For example, in the Mediterranean basin, sensitivity to land degradation (Fig. 2.6) is generally linked to the ecological conditions typical of the place (e.g. climatic aridity, soil characteristics, erosion, slope and vegetation cover) in combination with certain aspects of anthropogenic pressure, such as population density, and unsustainable use of the soil resource (Mensching 1986; Moonen et al. 2002; Sivakumar 2007).

The sensitivity to land degradation in this geographical region has attracted interest from researchers (e.g. Puigdefábregas and Mendizabal 1998) and not only, but also considering the effects it can have on food and energy security, water availability and on the ability to adapt to climate change. Furthermore, as mentioned previously, the phenomenon usually involves a temporary reduction or loss of the biological and economic productivity of fertile soils, pastures and wooded areas (Fantechi et al. 1995; Tanrivermis 2003; Salvati and Zitti 2009). For this reason, some authors (Marathianou et al. 2000; Máñez Costa et al. 2011; Brevik et al. 2020) underline how land degradation is to be considered a serious threat to agriculture and agricultural communities, and even human health.

If the traditional biophysical approach suggests that certain processes and systems of the terrestrial ecosystem have been lost due to improper use and abuse of the soil by humans, a socio-economic approach appears increasingly important for the understanding of the key factors and outcomes of the degradation processes underway in

---

the need to specify the meanings associated with "SENSITIVITY" and "VULNERABILITY". The first term should be used to indicate the "level of degradation reached in a certain territory by the processes of desertification due to climate change, soil erosion, deforestation, salinization, etc. triggered by natural or anthropic causes"; the second should instead express the "level of susceptibility (or resilience) to the phenomena of desertification".

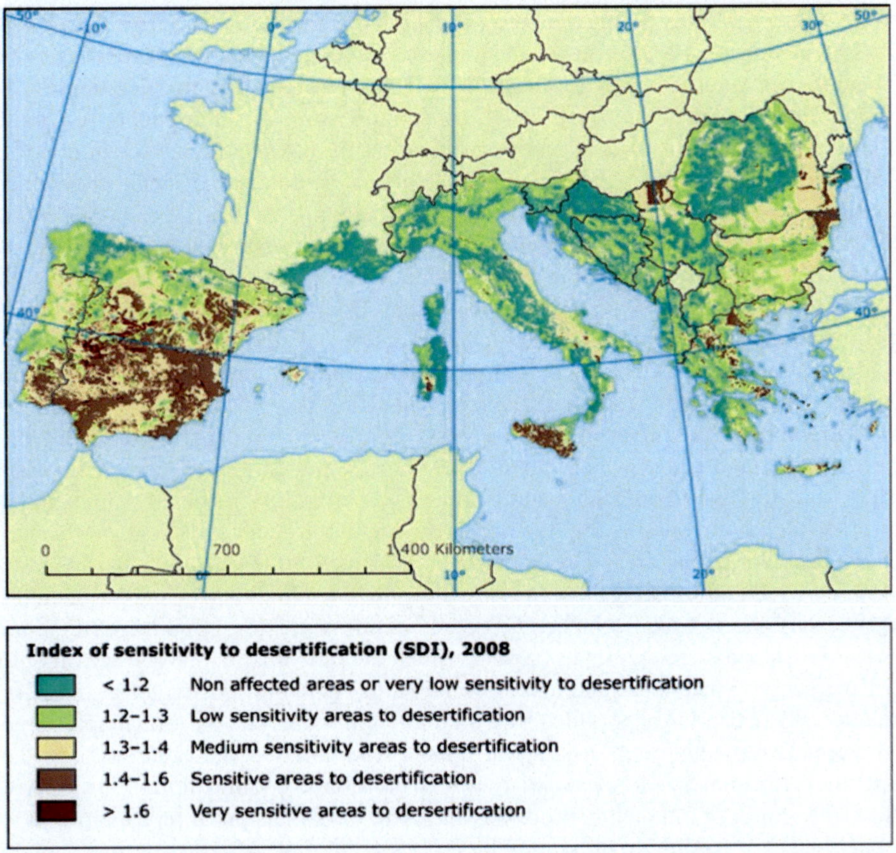

Fig. 2.6 Map of the DISMED (Desertification Information System for the Mediterranean) project showing sensitivity to land degradation in southern Europe. *Source* Website of the European Environment Agency

recent decades, highlighting the relationships between land use and land management practices (Salvati and Zambon 2018).

Land degradation represents one of the most current and important environmental issues that our societies are called to face, due to the serious consequences it poses to the health of the environment and man.[17] If, on the one hand, the issue has often been at the center of the attention of the media, decision-makers and public opinion, on the

[17] As well expressed by Rapport et al. (1998), the notion of "health" has generally been used to denote the vitality of individuals and, more recently, of populations (humans, domestic animals and wild animals). The extension of the term "health" to describe regional ecosystems is a response to evidence of experiences showing that certain man-managed ecosystems have become highly dysfunctional. Extending the notion of health to the regional scale (ecosystems, river basins, landscapes) offers new opportunities to integrate social, natural and health sciences. A healthy ecosystem is defined as "stable and sustainable"; it is able to maintain its organization and autonomy over time and its resistance to stress.

other, the tangentially cyclical nature of this interest must be noted, corresponding, in its peak phases. The emergence of situations associated with the sphere of climate change is also remarkable (e.g. Safriel 2007). This interest has directed the attention of the public toward the desertification–climate relationship (and more generally toward the biophysical factors that underlie desertification, the last stage of land degradation), leading instead to neglect the important role played by contextual social, economic, cultural, political and institutional factors (Dematteis 1998; Di Feliciantonio et al. 2018). The problem of land degradation, even before the irreversible phenomena of desertification, would instead require continuous and systematic attention through the permanent monitoring of a phenomenon that strongly affects the quality of life of people and the stability of ecosystem balance and which, therefore, to be effectively addressed (Perrin et al. 2018), it requires the development of calibrated medium and long-term intervention strategies (Lamonica et al. 2020).

Different evaluation methods have been developed for the description of the "degraded" landscape, which often rely on the integration of numerous variables (quantitative approaches) and synthetic vulnerability indices (computational approaches). For example, the use of GIS methodologies and statistical approaches is widespread in this area (e.g. Shalaby and Tateishi 2007; Rahman et al. 2009). A study by Incerti et al. (2007) can be exemplified in this sense which, to identify the regions characterized by a certain degree of climatic aridity, and therefore more sensitive to land degradation, in Italy, classified the bioclimatic time series available for the geographical area under an artificial neural network.[18] Another example of the use of time series is provided by a study by Hill et al. (2008), which applies an interpretative methodology to map the spatial extent of specific syndromes, in the Iberian Peninsula. Syndromes, which we can consider a combination of symptoms, describe groups of interacting processes and symptoms that appear repeatedly and in many places according to characteristic combinations and patterns. In the example considered, we are talking about over-exploitation (caused by excessive grazing and the extraction of timber), particular disasters (connected to fires) and abandonment of agriculture (linked to the spontaneous spread of vegetation) that are put in relation to the temporal trend of an indicator of plant productivity (the NDVI, acronym for Normalized Difference Vegetation Index).

As detectable by numerous scientific contributions (e.g. De Paola et al. 2013; Luca Salvati and Ferrara 2015; Symeonakis et al. 2016; Mina Karamesouti et al. 2018), the ESA method is widely used especially in the Mediterranean area, which integrates environmental indicators produced through the computation of data from

[18] Artificial neural networks, objects of great interest for the application models that derive from them in the scientific world, arise from the idea of simulating some functions and capabilities of the human brain thus giving the machine a certain decision-making capacity thanks to its "artificial intelligence". An artificial neural network is made up of a large number of independent units, connected by means of links, through which impulses pass. It usually works through a certain algorithm that modifies the weights (attenuations) of the connections, so that it adapts to provide a certain output in response to a certain input: this is the "training" procedure of the network in which it is instructed to perform a certain task (Zell et al. 1994).

multiple information sources. These relate to the three biophysical dimensions—climate, soil, vegetation—which directly influence soil degradation processes. A synthetic index of vulnerability to land degradation, called ESAI (Environmental Sensitive Area Index), falls within this methodological framework, which considers together numerous variables and thematic indicators concerning the climate, soil quality, plant cover and soil management. ESAI has been validated both locally and regionally in many areas of southern Europe (for example Portugal, Spain, Italy, Greece) under different environmental conditions (Basso et al. 2000). For example, Contador et al. (2009) have shown that ESAI is well correlated with many independent variables associated with soil degradation in a large region of Spain and is therefore suitable for correctly identifying areas potentially affected by land degradation. Through the computation of this index, Salvati and Zitti (2009) have identified regional convergences of environmental variables in Italy, providing useful information to guide policies regarding complex processes characterized by the interaction of ecological and economic factors.

In some cases, the socio-economic factors are those that play a greater role in the manifestation of land degradation, for example in the case of indiscriminate urbanization (Prasad and Badarinth 2004). The interaction between the "environmental" and "socio-economic" components of the territorial systems that show some form of land degradation is multifaceted and changeable over time (Ibáñez et al. 2008). This evolution implies changes in time and space of multiple interacting factors, some of which can be governed by man, others not. Based on the hypothesis that ecological and socio-economic factors contribute differently to land degradation, creating a complex spatial and temporal pattern of environmental degradation. Salvati 2014a) sought to interpret the growing disparities in soil vulnerability because of the socio-economic dynamics that contribute to unsustainable development paths. Their analysis confirmed that poor land management triggered by unsustainable socio-economic development is emerging as a crucial aspect of land degradation in areas that are already vulnerable.

Evaluating the health of the ecosystem in relation to the ecological, economic and human health sphere requires the integration of human values with biophysical processes, an integration that, even from what emerges from this brief review, is difficult to find in conventional science. Due to the complexity and multi-dimensionality of the phenomenon, the operational definition, long-term monitoring and study of land degradation represent a complex task (Dallara and Rizzi 2012), but which is of fundamental importance, especially in a perspective of sustainability. This can be achieved through an analysis of the relationships between the pressures that human places on ecosystems and the landscape, the structure and function of altered ecosystems, the alteration of ecosystem services and the effects on society.

Fig. 2.7 Soil sealing by its very nature strongly affects the soil, decreasing many of its beneficial effects

2.2.3 Why Land Degradation and Soil Sealing Are Connected?

The debate on land degradation neutrality is strongly linked to the question of which functions and threats to soil are most important in a given geographical area. The main threats have been recognized at European level through the Communication from the Commission to the European Parliament and Council "Towards a Thematic Strategy for Soil Protection" (European Comission 2002). Among the eight soil degradation processes mentioned, there is SOIL SEALING (also called, in Italian, soil waterproofing),[19] a term by which we mean the destruction or covering of the soil by buildings or other structures made up of partially or partially completely waterproof, resulting in biophysical degradation. This phenomenon leads to the complete loss of soil functions and is usually irreversible.

Soil sealing strongly reduces the possibility of soil to perform its functions (Fig. 2.7) by interrupting the cycle of water, nutrients and biological cycles (Munafò et al. 2013; Bimonte and Stabile 2019).

It is a common practice to remove the upper arable layer, which provides most of the services connected to the ecosystem, to develop strong foundations in the subsoil and/or in the underlying rock that support the building or infrastructure before proceeding with the rest of the construction. This separates the pedosphere from the atmosphere, preventing the infiltration of rain and the exchange of gas between soil and air. As a result, waterproofing consumes the soil (unless you properly reuse it elsewhere). This is of serious concern, given that the soil takes a long time to form and it takes centuries for even one centimeter to form (Alexander 1988).

[19] Soil sealing, erosion, compaction, and loss of organic carbon, contamination, excessive fertilization, loss of biodiversity, desertification, salinization, acidification, and landslide.

Among the main impacts of soil waterproofing is the strong pressure exerted on water resources, which causes changes in the environmental status of the water collection basins, affecting the ecosystems and the services they offer. Waterproofing reduces the absorption of rain in the soil, in extreme cases, preventing it completely (Singer and Le Bissonais 1998). The infiltration of rainwater into soils sometimes causes it to take longer to reach rivers, reducing the flow and therefore the risk of floods. Much of the water resources in the soil are absorbed by the plants, reducing the incidence of drought and therefore avoiding the need for irrigation, with fewer problems of salinization in agriculture. In addition, greater infiltration of water reduces the dependence on artificial plants (for example from a basin) for the collection of precipitation peaks. In this way, the capacity of the soil and of the vegetation that grows there, to retain water allows to store it instead of collecting, channeling and purifying the drain. On the contrary, in cities where the soil is highly waterproofed, the capacity of the sewage system may not be sufficient for the high water flow, causing surface flooding.

The increase in the solid contribution of the flowing waters and the polluting load of the same causes, among other things, a strong impact on the quality of surface waters and therefore also on aquatic life. Above all, the reduction of vegetation and the removal of the surface layer rich in an organic substance due to soil sealing compromise its production functions, as well as the possibility of absorbing CO_2 and providing support and sustenance for the biotic component of the ecosystem. A waterproofed ground also contributes to making the climate warmer and drier due to the lower evapotranspiration and the larger surfaces with a high heat refractive coefficient (Tombolini et al. 2015). Finally, the migratory corridors for wild species are interrupted and the fragmentation of habitats is increased, reducing their biodiversity (EEA 2004).

The local waterproofing patterns are not uniform. The degree to which an area is affected by soil sealing is related not only to the different types of land use but also to the time when the territory was artificialized and urban planning. In this regard, it is interesting to note that urban soils are characterized by strong spatial and temporal heterogeneity (Effland and Pouyat 1997). Firstly, because land use in urban contexts can vary considerably over time (for example, from industrial to residential, public or recreational); secondly, the numerous inputs of exogenous material and the mix of this with the original lithological material leads to high spatial variability. Generally, the most modified soil is close to the urban core.

Built-up areas represent around 4% of the territory monitored by the European Environment Agency (Jones et al. 2012). Despite the objective of slowing down the waterproofing of the soil, the spread of built areas continues to increase throughout Europe and especially in the western one.

Driven mainly by urban sprawl (Scalenghe and Marsan 2009), the waterproofing of soils with good land quality is considered a significant threat for European countries. Changes in land use and its consumption can have unpredictable consequences for ecologically fragile landscapes, determining or consolidating disparities in access to soils with high quality (Ferrara et al. 2014a).

Even in the context of the program to achieve land degradation neutrality, it has been recognized that in the western world the change in land use is the factor that most influences land degradation and represents one of the three indicators considered universal to measure this phenomenon.[20] In the conceptual methodological framework for a land degradation neutrality, the conversion of forests, natural and productive areas into settlements are considered universally negative, causing land degradation.

2.2.4 *The Worn Landscape*

The progressive spread of artificial surfaces and, therefore, the cementing of the territory, linked to the settlement dynamics and the expansion of urban residential areas and related infrastructures, is responsible for the consumption of soil.

SOIL CONSUMPTION is a phenomenon associated with the loss of a fundamental environmental resource, due to the occupation of originally agricultural, natural or semi-natural surface, and is defined as a variation from a non-artificial cover (unconsumed soil) to a cover artificial soil (worn soil). Land use (i.e. the new artificial cover) can be divided into two main categories.

- Permanent soil consumption: buildings, buildings; asphalted roads; railway headquarters; airports (runways and waterproof/paved handling areas); ports (docks and waterproof/paved handling areas); other waterproof/paved areas; permanent paved greenhouses; and, landfills.
- Reversible soil consumption: dirt roads; construction sites and other areas in beaten earth (squares, parking lots, courtyards, sports fields, permanent material storage); non-renaturalized extraction areas; quarries in the pitch; photovoltaic fields on the ground; and, other artificial coverings whose removal restores the initial soil conditions.

One of the modes of growth of urban areas with which land use is associated is manifested in the forms of expansion of the edge of the urban center, with the creation of new neighborhoods or residential areas that maintain the main characteristics of the urban core that are going to expand (Munafò et al. 2011). The current expansion of waterproofed surfaces is largely attributable to unplanned urban development, which in the urban and peri-urban fringe of many important cities mixes different types of land use (European Environment l Agency 2006). The spread of low density dispersed settlements from the urban center to the outside, known with the English term URBAN SPRAWL, is a well-studied traditional phenomenon for cities in North America (Bruegmann 2006; Duany et al. 2010) in place since the beginning of the twentieth century, while it is relatively recent as regards European cities (Kasanko et al. 2006; Couch et al. 2008; Salvati and Zitti 2011). As will be seen in more detail in the following chapter, many urban areas of the Mediterranean region have also

[20] The other two indicators are net primary productivity and carbon stock (Orr et al. 2017).

gradually lost their compactness, evolving toward a more widespread structure, due to a changed form of the expansion of residential and commercial settlements, and related infrastructures (Leontidou 1990; Catalán et al. 2008; Luca Salvati and Morelli 2014). This produced a subtle process of simplification and deterioration of the rural landscape, due to the reduction or disappearance of the arable land and vineyards (Polyzos et al. 2008; Morelli et al. 2014; Ciommi et al. 2019).

The two types of urban growth—compact and dispersed—have very different soil consumption characteristics. If the compact growth involves an almost total saturation and waterproofing of the natural soil, the widespread city is instead characterized by the coexistence of buildings and green areas, which guarantee a partial persistence of the natural characteristics of the soils concerned, at least for the portions on which they do not insist built buildings directly. The two types also differ in the degree of compromise they exert outside urbanized areas: while the impacts on the peri-urban fringe areas due to the "compact" urban extensions are relatively limited, in the case of sprawl, a substantial part of the margin surfaces urban is removed from the original intended use, due to the fragmentation and transformation of the spaces, of which both agricultural productivity and natural characteristics are compromised (Terzi and Bolen 2009; Frondoni et al. 2011; Munafò et al. 2011).

2.3 The Case of Italy

The two most evident phenomena of the metamorphosis of the Italian landscape, the expansion of urban areas and the renaturalization following the abandonment of crops, are the consequence of the great social and economic changes that have taken place since the Second World War to the present day. From the second half of the twentieth century, thanks also to impressive economic growth and the evolution of the transport system with important investments in infrastructure and private mobility, we have witnessed the gradual transition from the centrality of agriculture to that of industry and services. This is how the decade 1950–1960 saw the beginning of a very rapid transformation in the expansion of the city, which, gradually invading the surrounding natural and agricultural free zones, came to occupy free spaces also very distant from the historical urban nuclei, often without any programming (Munafò and Marinosci 2018).

The pattern of the 1960s configures a dipolar Italy, with an evident urban–rural gradient and a moderate dispersion of settlements, concentrated mainly around large urban centers. The classic polarities have gradually disappeared over the decades, while an opaque scenario of latent and widespread urbanization has invaded the coastal areas (e.g. Falco 2017) extending also toward the internal regions (e.g. Tuscany, Calabria, Abruzzo and Basilicata; (Romano and Zullo 2014). The disintegration of monocentric models, the formation of large urban voids accompanied the urbanization of the settlement areas, in Milan as in Rome, in Naples as in Turin

(Carlucci et al. 2017). Urban centers have become saturated, reaching soil waterproofing rates much higher than 50%. The Po Valley has been also hit by a fragmented urban model, gradually losing its rural character. Mountain areas were not spared either: moving from a strict conservation regime, exercised in protected areas, to a progressive diversification of more accessible landscapes, a dilapidated local development produced the coexistence of low-density productive residential areas with fragmented agricultural areas and discontinuous, with a high per capita soil consumption (Zambon et al. 2017b).

The widespread city is now a characteristic element of vast areas of the Italian territory, where urban sprawl tends to eliminate the distinction between city and countryside with high social, economic and environmental costs, without being accompanied by real housing needs. If, in fact, the compact urban expansions, a typical expression of the building boom that in the sixties and seventies determined the transformation and growth of the main urban centers, had demographic dynamics as determining factors. The development of the widespread city typical of the most recent decades is mainly due to the cultural transformation of the expectations of the population in relation to factors related to the quality of life and, in particular, to the spread on a large scale of economic well-being. The choice to live even at considerable distances from the urban center is supported, inter alia, by the development of transport infrastructures that allow fast daily movements between the home and the workplace or study. Other factors that determine the increase in surfaces intended for residences in peri-urban areas are attributable to the lower costs of the real estate units (Salvati et al. 2012).

Due to the peculiar topographical conditions, due to the wide availability of building land in the Roman countryside, Rome was for many years an example of a less compact urban area compared to other Mediterranean cities (such as Naples, Barcelona and Athens) and with a particularly chaotic, fragmented and somewhat polycentric development, to the point that Fratini (2000) defines the recent urban development of the capital of Italy with the expression "an archipelago of urban islands". The insularization and fragmentation of the built-up area have long represented the urban tendencies of the entire metropolitan area, originally characterized by a fine plot of rather compact and orderly rural settlements around the central core. In the 60s and 80s, urban sprawl occurred, together with an equally strong repolarization phase, both in the central area of the Grande Raccordo Anulare (GRA) and in the more marginal areas and around the smaller centers. In recent decades, there has been a development of the metropolitan area through ex-Novo expansion processes on the one hand (Figs. 2.8 and 2.9) and, on the other, densification processes in both areas of the GRA and many peripheral municipalities of the internal hill.

The updated data on land use in Italy have been produced annually with detail on a national, regional and municipal scale, thanks to the institutional activities of the National System for Environmental Protection (SNPA), which sees the Higher Institute for Protection and Environmental Research (ISPRA) together with the Environmental Protection Agencies of the Regions and Autonomous Provinces (ARPA/APPA) in joint monitoring work, carried out using the best information that new technologies can offer, such as those deriving from earth observation satellites.

Fig. 2.8 Rome Eur 2001 and 2017 (Google Earth images)

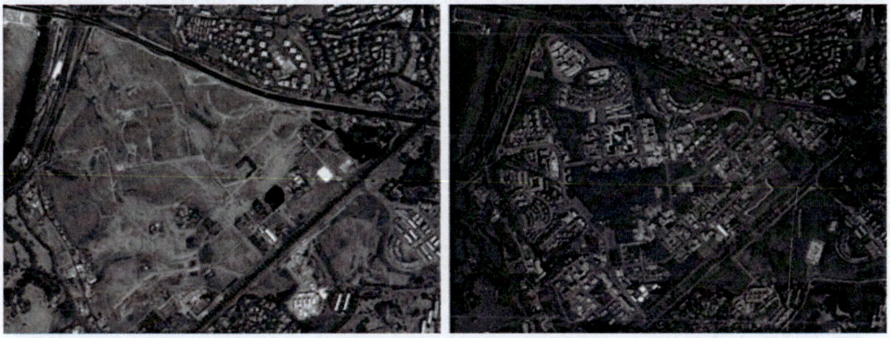

Fig. 2.9 Rome Torrino Mezzocammino 2001 and 2017 (Google Earth images)

Land use in Italy averaged 8 m^2 per second, from 2.8% in 1956 to 6.9% in 2010 (cementing more than 20,500 km^2 of land; Munafò et al. 2013), up to 7, 65% in 2017 according to the new data provided by the National Environmental Protection System and ISPRA. The built-up areas are not homogeneous in terms of size and use, with a combination of rural, residential, industrial and tertiary functions (building density: 0.1 buildings per hectare, population density: 0.2–0.5 inhabitants per hectare). The changes that have occurred in urban structures have affected Italy in different ways, not always according to the number of the population (Munafò and Tombolini 2014): since the 1980s the trend of the resident population has tended to flatten, contrary to the phenomenon of waterproofing soil that never seems to know setbacks. This phenomenon therefore cannot be explained simply by population growth, nor by the potential housing demand and urbanization of the territory connected to it (Salvati et al. 2016).

> We consume soil by losing citizens or for 'ghost citizens': an even more useless, dissipative cementing. Urban planning policies which are likely to have generated two situations: a) the new building remains empty; b) the new building was filled by emptying part of the old which in the meantime has become abandoned or unused, as are many internal areas or many historic centers or as production areas on the edge of the urban area. Two different sides

> of a single major cultural, political and urban breakdown which still does not understand the serious situation it has before and against which it is necessary to react with decidedly alternative policies to those used to date.[21]

So says the consumption of soil in Italy Paolo Pileri, from the Polytechnic of Milan, calling it "persistent and inefficient", underlining how all this occurs even though in recent years some regions have adopted regional laws and the Chamber of Deputies approved in May 2016 a bill for soil protection (Senate Act No. 2383).[22]

The latest data show how land use in Italy continues to grow, albeit at a slower rate than that achieved a decade ago. During 2017, the new artificial coverages concerned 54 square kilometers of territory, or, on average, about 15 hectares per day, with a net growth of 52 square kilometers. The slowdown in the rate of land use compared to the beginning of the decade 2000–2010, surely due to the persistence of an economic crisis that heavily involved the world of construction and construction, seems to be in its terminal phase. In particular, in some regions, there is a first trend reversal, with a progressive artificialization of the territory, which increases in intensity, with the recovery of the building sector, the expansion of urban areas, often with low density, the densification of the existing fabrics and the launch of new construction sites for infrastructures and other works.

The most affected areas are the plains of the north, the Tuscan axis between Florence and Pisa, Lazio, Campania and Salento, the main metropolitan areas, the coastal strips, in particular, the Adriatic, Ligurian, Campania and Sicilian ones. In 15 regions, 5% of land consumption is exceeded, with the highest percentage value in Lombardy (which reaches 13%), in Veneto (over 12%) and Campania (over 10%). Emilia-Romagna, Friuli Venezia Giulia, Lazio, Puglia and Liguria follow, with values between 8 and 10%. Valle d'Aosta is the only region that has remained below the 3% threshold. Lombardy also holds the record in absolute terms, exceeding 310 thousand hectares of its artificially covered territory this year (13.4% of Italian artificial areas are in this region), compared to 9,500 hectares in the Aosta Valley.

The territorial distribution of the urbanized areas in Italy, therefore, sees the greatest densities located in correspondence with the main metropolitan areas. In large urban areas, there is a phenomenon of transformation caused by the movement of the inhabitants of the center toward new peripheral areas, in search of greater housing quality and different building and urban types, a phenomenon also driven by the localization pressure of the accommodation facilities and recreational areas in the areas with the greatest tourist flow.

At the provincial level, Monza and Brianza were, as in the previous year, the province with the highest percentage of artificial soil, with about 41% of soil consumed and a further significant increase of 35 hectares. Over 20% are the

[21] Text taken from the report of the National System for Environmental Protection—SNPA, of 2017, "Land use, territorial dynamics and ecosystem services", p. 160.

[22] Strong criticisms came, however, to the final text which, according to many, was not very effective and not able to ensure a real containment of land consumption due to the numerous derogations provided, the complex procedure for defining the limits and the fact that the reduction percentages to be reached over the years until 2050 were not established. At the beginning of this term, some bills were presented which, in part, take up and update the previous text.

provinces of Naples (34%), Milan (32%), Trieste (23%) and Varese (22%) and, slightly below, Padua (19%) and Treviso (17%). The only provinces that remained below the 3% threshold were Verbano-Cusio-Ossola (2.85%), Matera (2.87%), Nuoro (2.89%) and Aosta (2.91%). Among the latter, only Aosta grew more than the national average (+0.29%).

The province where the largest percentage increase in land use occurred compared to 2016 is that of Viterbo (+0.91%) (Fig. 2.10), followed by Verona (+0.71%), Vicenza (+0.67%), Bolzano (+0.65%), Venice (+0.57%), Vercelli (+0.54%) and Treviso (+0.49%).

The provinces of Isernia, Savona, Lucca, Massa Carrara, La Spezia, Caltanissetta and Cosenza are those, conversely, where the net percentage growth has been less. In the provinces of Veneto and Bolzano, the high increase corresponds to a net land consumption greater than 200 hectares in absolute terms. The record for the last year is in Verona, which touches the 300 hectares of new artificial soil, followed by Vicenza (+239). In absolute terms, the province of Rome is the only one to cross the 70,000-hectare threshold, followed by Turin (about 60,000 hectares), Brescia (more than 55,000 hectares) and Milan (which stands at 50,000 hectares).

More than a fifth (21.4%, almost 5,000 km^2) of artificial soil in Italy, in 2017, is concentrated in the territory administered by the 14 metropolitan cities; beyond the major metropolitan areas, in Northern Italy, many provinces have percentages of land consumption above the national average, together with other coastal areas of Tuscany, Lazio, Campania and Marche and, above all, Puglia and south of Sicily.

The orographic conformation of the territory heavily affects the geography of urbanization, which is concentrated in the foothills (such as the Lombard-Venetian one), in the plains and coastal areas. Of particular concern is the intense urbanization of the coasts, which, almost seamlessly, covers the Adriatic coast, but also of long stretches of the Tyrrhenian, Ionian and Islands (Salvati and Carlucci 2016).

The percentage values of the soil consumed, therefore, increase as they approach the coast (Table 2.8). At a national level, almost a quarter of the strip within 300 m from the sea is now worn out.

Confirming the 2016 data, among the regions with the highest recorded values within 300 m from the coastline are Liguria and Marche with almost 50% of the land consumed, Abruzzo, Campania, Emilia-Romagna and Lazio with values between 30 and 40%. Between 300 and 1,000 m, Abruzzo, Emilia-Romagna, Campania, Liguria and Marche are worth mentioning with values equal to or greater than 30% of consumption.

In the range between 1 and 10 kms, the figure of 16.4% of the consumption is evident, representative of the Campania Region. Once again, the largest percentage increase between 2016 and 2017 is recorded in the range between 1 and 10 kms from the coast: the increase in Friuli Venezia Giulia (+1.13%) compared to the data collected in 2016 On the other hand, the increase is more contained, in the areas adjacent to the sea, where the level of land consumption has now left a minimum quantity of undeveloped areas outside the protected areas. Nonetheless, construction continues also in the area below 300 m, with an increase in the land consumed by 0.10% nationally.

Fig. 2.10 Viterbo, area of the Murialdo district. In recent years and, in particular since 2013, the Murialdo district has seen the development of a new residential complex to replace an area intended for recreational activities which is part of the parish of S. Leonardo Murialdo and already existed in 1968. Like many other cities Viterbo has been affected, starting from the sixties, by a considerable shift of the inhabited nucleus towards the countryside. Furthermore, even if due to the flattening of the photo it is not perceived, it is probable (at least we get this impression from the reconstructions from memory of the landscape, and from details of the photo) that, during the construction of some of the blocks, the areas originally cultivated with olive trees, for a total of the order of tens of thousands of square meters, have been lowered by a few meters. Above: Photograph dating back to August 1968 [Annunzio Celaschi, with a non-professional camera, about 300 m altitude]. Bottom: image [with 3d option] Google Earth, 2017

Table 2.8 Percentage of land consumption compared to the distance from the coastline on a regional basis, excluding regions that are not washed by the sea (2017) and percentage increase compared to 2016

Region	Within 300 m		Between 300 and 1000 m		Between 1 and 10 km		Over 10 km	
Veneto	11.2	+0.21	10.8	+0.50	13.2	+0.45	10.7	+0.52
Friuli Venezia Giulia	13.7	+0.36	14.3	+0.08	13.8	+1.13	7.2	+0.31
Liguria	48.1	+0.05	31.0	+0,06	9.2	+0.07	2.2	+0.01
Emilia Romagna	34.2	+0.00	31.9	+0.10	12.7	+0,13	9.0	+0.22
Tuscany	21.5	+0.00	16.6	+0.16	9.5	+0.11	5.7	+0.10
Marche	45.7	+0.04	30.0	+0.07	12.0	+0,20	4.7	+0.27
Lazio	31.2	+0.05	21.7	+0.11	11.1	+0.38	6.4	+0.19
Abruzzo	36.6	0.18	31.8	+0.12	11.1	+0.31	3.5	+0.19
Molise	19.9	+0.00	16.5	+0.11	5.2	+0.15	3.5	+0.22
Campania	35.1	+0.09	30.2	+0.08	16.4	+0.16	6.5	+0.23
Puglia	29.8	+0.15	21.8	+0.24	10.3	+0.25	4.3	+0.27
Basilicata	6.0	+0.11	5.0	+0.62	5.4	+0.03	3.1	+0.11
Calabria	29.4	+0.07	20.1	+0.09	5.1	+0.07	2.1	+0,06
Sicily	28.8	+0,13	24.8	+0.14	10.6	+0,20	2.8	+0.11
Sardinia	10.4	+0.04	8.8	+0.27	4.9	+0,20	1.8	+0.03
Italy	**23.4**	**+0.10**	**19.6**	**+0.16**	**9.3**	**+0.23**	**7.0**	**+0.23**

Source ISPRA elaborations on SNPA cartography

Altimetry significantly affects soil consumption: about 11.9% of the soil is consumed at an altitude of fewer than 300 m a.s.l. The percentage of consumed tends to decrease with increasing altitude, reaching 5.8% between 300 and 600 m, and only 2.7% over 600 m.

The main communication routes also represent privileged axes for urban development. The development of large linear infrastructures shows a concentration especially in Northern Italy, where the volume of traffic is greater. However, it is the entire national territory, which, from the coastal dunes to the alpine glaciers, is furrowed by a dense network that branches off for kilometers without interruption. The infrastructure is a key factor of territorial development, however, its development should have been planned with attention to the effects on the settlement system (attraction and increase of urbanized areas) but also concerning the negative influence on the quality of the landscape and environment, determined by the fragmentation of the landscape with insurmountable barriers for animal and plant species that can no longer find a suitable habitat for their survival.

References

Abu Hammad A, Tumeizi A (2012) Land degradation: socioeconomic and environmental causes and consequences in the Eastern Mediterranean. Land Degrad Dev 23:216–226. https://doi.org/10.1002/ldr.1069

Akhtar-Schuster M, Stringer LC, Erlewein A, Metternicht G, Minelli S, Safriel U, Sommer S (2017) Unpacking the concept of land degradation neutrality and addressing its operation through the Rio Conventions. J Environ Manage 195:4–15. https://doi.org/10.1016/j.jenvman.2016.09.044

Alexander EB (1988) Rates of soil formation: implications for soil loss tolerance. Soil Sci 145:37–45

Bajocco S, De Angelis A, Salvati L (2012) A satellite-based green index as a proxy for vegetation cover quality in a Mediterranean region. Ecol Ind 23:578–587. https://doi.org/10.1016/j.ecolind.2012.05.013

Bajocco S, Salvati L, Ricotta C (2011) Land degradation versus fire: a spiral process? Prog Phys Geogr: Earth Environ 35:3–18. https://doi.org/10.1177/0309133310380768

Basso F, Bove E, Dumontet S, Ferrara A, Pisante M, Quaranta G, Taberner M (2000) Evaluating environmental sensitivity at the basin scale through the use of geographic information systems and remotely sensed data: An example covering the Agri basin (Southern Italy). CATENA 40:19–35. https://doi.org/10.1016/S0341-8162(99)00062-4

Biasi R, Brunori E, Smiraglia D, Salvati L (2015) Linking traditional tree-crop landscapes and agro-biodiversity in Central Italy using a database of typical and traditional products: a multiple risk assessment through a data mining analysis. Biodivers Conserv 24(12):3009–3031

Biasi R, Colantoni A, Ferrara C, Ranalli F, Salvati L (2015) In-between sprawl and fires: long-term forest expansion and settlement dynamics at the wildland–urban interface in Rome, Italy. Int J Sust Dev World 22:467–475. https://doi.org/10.1080/13504509.2015.1064488

Bimonte S, Stabile A (2019) The effect of growth and corruption on soil sealing in Italy: a regional environmental Kuznets Curve Analysis. Environ Resource Econ 74:1497–1518. https://doi.org/10.1007/s10640-019-00376-1

Boellstorff D, Benito B (2005) Impacts of set-aside policy on the risk of soil erosion in central Spain. Agr Ecosyst Environ 107:231–243

Bogaert J, Barima YSS, Mongo LIW, Bamba I, Mama A, Toyi M, Lafortezza R (2011) Forest fragmentation: causes, ecological impacts and implications for landscape management. In: Li C, Lafortezza R, Chen J (eds) Landscape ecology in forest management and conservation: challenges and solutions for global change. Springer, Berlin, Heidelberg, pp 273–296. https://doi.org/10.1007/978-3-642-12754-0_12

Bouma J (2002) Land quality indicators of sustainable land management across scales. Agric Ecosyst Environ 88. Soil Health as an Indicator of Sustainable Management: 129–136. https://doi.org/10.1016/S0167-8809(01)00248-1

Brevik EC, Slaughter L, Singh BR, Steffan JJ, Collier D, Barnhart P, Pereira P (2020) Soil and human health: current status and future needs. Air, Soil and Water Res 13. https://doi.org/10.1177/1178622120934441

Bruegmann R (2006) Sprawl: a compact history. University of Chicago Press

Canfora L, Salvati L, Benedetti A, Francaviglia R (2017) Is soil microbial diversity affected by soil and groundwater salinity? Evidences from a coastal system in central Italy. Environ Monit Assess 189:319. https://doi.org/10.1007/s10661-017-6040-1

Carlucci M, Chelli FM, Salvati L (2018) Toward a new cycle: short-term population dynamics, gentrification, and Re-Urbanization of Milan (Italy). Sustainability 10:1–12

Carlucci M, Grigoriadis E, Rontos K, Salvati L (2017) Revisiting a hegemonic concept: long-term 'Mediterranean Urbanization' in between city re-polarization and Metropolitan Decline. Appl Spat Anal Policy 10:347–362. https://doi.org/10.1007/s12061-016-9186-2

Carter MR, Gregorich EG, Anderson DW, Doran JW, Janzen HH, Pierce FJ (1997) Concepts of soil quality and their significance. In: Gregorich EG, Carter M (eds) Soil quality for crop production and ecosystem health. Elsevier Science Publishers, Ámsterdam, Países Bajos, pp 1–20

Catalán B, Saurí D, Serra P (2008) Urban sprawl in the Mediterranean? Patterns of growth and change in the Barcelona Metropolitan Region 1993–2000. Landscape Urban Plan 85:174–184

Cecchini M, Zambon I, Pontrandolfi A, Turco R, Colantoni A, Mavrakis A, Salvati L (2019) Urban sprawl and the 'olive' landscape: sustainable land management for 'crisis' cities. GeoJournal 84:237–255. https://doi.org/10.1007/s10708-018-9848-5

Chasek P, Safriel U, Shikongo S, Fuhrman VF (2015) Operationalizing Zero Net Land Degradation: the next stage in international efforts to combat desertification? J Arid Environ 112:5–13. https://doi.org/10.1016/j.jaridenv.2014.05.020

Ciommi M, Chelli FM, Carlucci M, Salvati L (2018) Urban growth and demographic dynamics in Southern Europe: toward a new statistical approach to regional science. Sustainability 10:2765. https://doi.org/10.3390/su10082765

Ciommi M, Chelli FM, Salvati L (2019) Integrating parametric and non-parametric multivariate analysis of urban growth and commuting patterns in a European metropolitan area. Qual Quant 53:957–979. https://doi.org/10.1007/s11135-018-0798-2

Ciommi M, Gentili A, Ermini B, Gigliarano C, Chelli FM, Gallegati M (2017a) Have your cake and eat it too: the well-being of the Italians (1861–2011). Soc Indic Res 134:473–509. https://doi.org/10.1007/s11205-016-1450-y

Ciommi M, Gigliarano C, Emili A, Taralli S, Chelli FM (2017b) A new class of composite indicators for measuring well-being at the local level: an application to the Equitable and Sustainable Well-being (BES) of the Italian Provinces. Ecol Ind 76:281–296. https://doi.org/10.1016/j.ecolind.2016.12.050

Cohen-Shacham E, Janzen C, Maginnis S, Walters G (2016) Nature-based solutions to address global societal challenges

Colantoni A, Ferrara C, Perini L, Salvati L (2015a) Assessing trends in climate aridity and vulnerability to soil degradation in Italy. Ecol Ind 48:599–604. https://doi.org/10.1016/j.ecolind.2014.09.031

Colantoni A, Grigoriadis E, Sateriano A, Venanzoni G, Salvati L (2016) Cities as selective land predators? A lesson on urban growth, deregulated planning and sprawl containment. Sci Total Environ 545–546:329–339. https://doi.org/10.1016/j.scitotenv.2015.11.170

Colantoni A, Mavrakis A, Sorgi T, Salvati L (2015b) Towards a 'polycentric' landscape? Reconnecting fragments into an integrated network of coastal forests in Rome. Rendiconti Lincei 26:615–624. https://doi.org/10.1007/s12210-015-0394-5

Contador JFL, Schnabel S, Gómez Gutiérrez A, Pulido Fernández M (2009) Mapping sensitivity to land degradation in Extremadura, SW Spain. Land Degrad Dev 20:129–144. https://doi.org/10.1002/ldr.884

Costanza R, d'Arge R, de Groot R, Farber S, Grasso M, Hannon B, Limburg K, Naeem S, O'Neill RV, Paruelo J (1998) The value of ecosystem services: putting the issues in perspective. Ecol Econ 25:67–72

Couch C, Petschel-Held G, Leontidou L (eds) (2008) Urban Sprawl in Europe: landscape, land-use change and policy. Oxford and Malden, MA

Dallara A, Rizzi P (2012) Geographic map of sustainability in Italian local systems. Reg Stud 46:321–337. https://doi.org/10.1080/00343404.2010.504703

De la Rosa D (2005) Soil quality evaluation and monitoring based on land evaluation. Land Degrad Dev 16:551–559

De Paola F, Ducci D, Giugni M (2013) Desertification and erosion sensitivity. A case study in southern Italy: the Tusciano River catchment. Environ Earth Sci 70:2179–2190. https://doi.org/10.1007/s12665-013-2294-2

Delfanti L, Colantoni A, Recanatesi F, Bencardino M, Sateriano A, Zambon I, Salvati L (2016) Solar plants, environmental degradation and local socioeconomic contexts: a case study in a Mediterranean country. Environ Impact Assess Rev 61:88–93. https://doi.org/10.1016/j.eiar.2016.07.003

Dematteis G (1998) Suburbanización y periurbanización. Ciudades anglosajonas y ciudades latinas. In: Monclús J (ed) La ciudad dispersa. Centre de Cultura Contemporània de Barcelona, Barcelona

Di Feliciantonio C, Salvati L, Sarantakou E, Rontos K (2018) Class diversification, economic growth and urban sprawl: evidences from a pre-crisis European city. Qual Quant 52(4):1501–1522
Duany A, Plater-Zyberk E, Speck J (2010) Suburban Nation: the rise of Sprawl and the decline of the American DREAM. New York
Duvernoy I, Zambon I, Sateriano A, Salvati L (2018) Pictures from the other side of the fringe: urban growth and peri-urban agriculture in a post-industrial city (Toulouse, France). J Rural Stud 57:25–35
EEA (2004) High Nature Value Farmland: Characteristics, trends and policy challenges. European Environment Agency, Copenhagen, Denmark
Effland WR, Pouyat RV (1997) The genesis, classification, and mapping of soils in urban areas. Urban Ecosyst 1:217–228. https://doi.org/10.1023/A:1018535813797
Esposito P, Patriarca F, Perini L, Salvati L (2016) Land degradation, economic growth and structural change: evidences from Italy. Environ Dev Sustain 18:431–448. https://doi.org/10.1007/s10668-015-9655-9
European Comission (2002) Communication from the Commission to the Council, the European Parliament, the Economic and Social Committee and the Committee of the Regions—Towards a Thematic Strategy for Soil Protection. European Union, Brussels, Belgium
European Environment Agency (200AD) Mapping sensitivity to desertification (DISMED). Data File. Copenhagen, Denmark
European Environment l Agency (2006) Urban sprawl in Europe—the ignored challenge. Publication. EEA Report No 10/2006. European Union, Copenhagen, Denmark
Falco E (2017) Protection of coastal areas in Italy: where do national landscape and urban planning legislation fail? Land Use Policy 66:80–89. https://doi.org/10.1016/j.landusepol.2017.04.038
Fantechi R, Peter D, Balabanis P, Rubio JL (1995) Desertification in a European context: physical and socio-economic aspects. Alicante, Spain
Ferrara A, Salvati L, Sabbi A, Colantoni A (2014a) Soil resources, land cover changes and rural areas: towards a spatial mismatch? Sci Total Environ 478:116–122. https://doi.org/10.1016/j.scitotenv.2014.01.040
Ferrara C, Salvati L, Tombolini I (2014b) An integrated evaluation of soil resource depletion from diachronic settlement maps and soil cartography in peri-urban Rome, Italy. Geoderma 232–234:394–405. https://doi.org/10.1016/j.geoderma.2014.05.020
Ferrara C, Carlucci M, Grigoriadis E, Corona P, Salvati L (2017) A comprehensive insight into the geography of forest cover in Italy: exploring the importance of socioeconomic local contexts. For Policy Econ 75. Special Section on The Economics of Carbon Sequestration in Forestry: 12–22. https://doi.org/10.1016/j.forpol.2016.11.008
Fini A, Frangi P, Mori J, Donzelli D, Ferrini F (2017) Nature based solutions to mitigate soil sealing in urban areas: results from a 4-year study comparing permeable, porous, and impermeable pavements. Environ Res 156:443–454. https://doi.org/10.1016/j.envres.2017.03.032
Forino G, Ciccarelli S, Bonamici S, Perini L, Salvati L (2015) Developmental policies, long-term land-use changes and the way towards soil degradation: evidence from Southern Italy. Scott Geogr J 131:123–140. https://doi.org/10.1080/14702541.2015.1047895
Forman RT, Forman RTT, Forman RTT (1995) Land Mosaics: the ecology of landscapes and regions. Cambridge University Press
Francaviglia R, Di Bene C, Farina R, Salvati L, Vicente-Vicente JL (2019) Assessing "4 per 1000" soil organic carbon storage rates under Mediterranean climate: a comprehensive data analysis. Mitig Adapt Strat Glob Change 24:795–818. https://doi.org/10.1007/s11027-018-9832-x
Fratini F (2000) Roma arcipelago di isole urbane. Uno scenario per il XXI secolo. Gangemi, Roma
Frondoni R, Mollo B, Capotorti G (2011) A landscape analysis of land cover change in the Municipality of Rome (Italy): Spatio-temporal characteristics and ecological implications of land cover transitions from 1954 to 2001. Landscape Urban Plan 100:117–128. https://doi.org/10.1016/j.landurbplan.2010.12.002
Gómez-Baggethun E, de Groot R, Lomas PL, Montes C (2010) The history of ecosystem services in economic theory and practice: from early notions to markets and payment schemes. Ecol Econ

69. Special Section—Payments for Environmental Services: Reconciling Theory and Practice: 1209–1218. https://doi.org/10.1016/j.ecolecon.2009.11.007

Gonzalez-Roglich M, Zvoleff A, Noon M, Liniger H, Fleiner R, Harari N, Garcia C (2019) Synergizing global tools to monitor progress towards land degradation neutrality: trends. Earth and the World Overview of Conservation Approaches and Technologies sustainable land management database. Environ Sci Policy 93:34–42. https://doi.org/10.1016/j.envsci.2018.12.019

Green A, Tzilivakis J, Warner DJ, Lewis KA (2017) Problems of benchmarking greenhouse gas emissions in dairy agriculture. Benchmarking Int J 24:1470–1489

Henderson BB, Gerber PJ, Hilinski TE, Falcucci A, Ojima DS, Salvatore M, Conant RT (2015) Greenhouse gas mitigation potential of the world's grazing lands: modeling soil carbon and nitrogen fluxes of mitigation practices. Agr Ecosyst Environ 207:91–100

Hill J, Stellmes M, Udelhoven Th, Röder A, Sommer S (2008) Mediterranean desertification and land degradation: mapping related land use change syndromes based on satellite observations. Glob Planet Chang 64. Climate Change and Desertification: 146–157. https://doi.org/10.1016/j.gloplacha.2008.10.005

Ibáñez J, Martínez J, Puigdefabregas J (2008) Assesing desertification risk using system stability condition analysis. Ecological Modelling (in press)

Incerti G, Feoli E, Salvati L, Brunetti A, Giovacchini A (2007) Analysis of bioclimatic time series and their neural network-based classification to characterise drought risk patterns in South Italy. Int J Biometeorol 51:253–263. https://doi.org/10.1007/s00484-006-0071-6

Jones A et al (2012) The state of soil in Europe: a contribution of the JRC to the European Environment Agency's environment state and outlook report—SOER 2010. JRC, Publications Office, Luxembourg

Kairis O, Karavitis C, Kounalaki A, Salvati L, Kosmas C (2013) The effect of land management practices on soil erosion and land desertification in an olive grove. Soil Use Manag 29:597–606. https://doi.org/10.1111/sum.12074

Kairis O, Karavitis C, Salvati L, Kounalaki A, Kosmas K (2015) Exploring the impact of overgrazing on soil erosion and land degradation in a dry mediterranean agro-forest landscape (Crete, Greece). Arid Land Res Manag 29:360–374. https://doi.org/10.1080/15324982.2014.968691

Kalantari Z, Ferreira CSS, Keesstra S, Destouni G (2018) Nature-based solutions for flood-drought risk mitigation in vulnerable urbanizing parts of East-Africa. Current Opinion in Environmental Science and Health 5:73–78. https://doi.org/10.1016/j.coesh.2018.06.003

Karamesouti M, Detsis V, Kounalaki A, Vasiliou P, Salvati L, Kosmas C (2015) Land-use and land degradation processes affecting soil resources: evidence from a traditional Mediterranean cropland (Greece). CATENA 132:45–55. https://doi.org/10.1016/j.catena.2015.04.010

Karamesouti M, Panagos P, Kosmas C (2018) Model-based spatio-temporal analysis of land desertification risk in Greece. CATENA 167:266–275. https://doi.org/10.1016/j.catena.2018.04.042

Karlen DL, Ditzler CA, Andrews SS (2003) Soil quality: why and how? Geoderma 114:145–156. https://doi.org/10.1016/S0016-7061(03)00039-9

Kasanko M, Barredo JI, Lavalle C, McCormick N, Demicheli L, Sagris V, Brezger A (2006) Are European cities becoming dispersed? A comparative analysis of 15 European urban areas. LandscAPE Urban Plan 77:111–130. https://doi.org/10.1016/j.landurbplan.2005.02.003

Keesstra S, Nunes J, Novara A, Finger D, Avelar D, Kalantari Z, Cerdà A (2018) The superior effect of nature based solutions in land management for enhancing ecosystem services. Sci Total Environ 610–611:997–1009. https://doi.org/10.1016/j.scitotenv.2017.08.077

Kelly C, Ferrara A, Wilson GA, Ripullone F, Nolè A, Harmer N, Salvati L (2015) Community resilience and land degradation in forest and shrubland socio-ecological systems: Evidence from Gorgoglione, Basilicata, Italy. Land Use Policy 46:11–20. https://doi.org/10.1016/j.landusepol.2015.01.026

Kertész Á, Madarász B (2014) Conservation Agriculture in Europe. Int Soil Water Conserv Res 2:91–96. https://doi.org/10.1016/S2095-6339(15)30016-2

Kosmas C, Danalatos NG, St Gerontidis (2000) The effect of land parameters on vegetation performance and degree of erosion under Mediterranean conditions. CATENA 40:3–17. https://doi.org/10.1016/S0341-8162(99)00061-2

Kosmas C, Detsis V, Karamesouti M, Kounalaki K, Vassiliou P, Salvati L (2015) Exploring long-term impact of grazing management on land degradation in the socio-ecological system of Asteroussia Mountains, Greece. Land 4:541–559. https://doi.org/10.3390/land4030541

Kosmas C, Karamesouti M, Kounalaki K, Detsis V, Vassiliou P, Salvati L (2016) Land degradation and long-term changes in agro-pastoral systems: an empirical analysis of ecological resilience in Asteroussia—Crete (Greece). CATENA 147:196–204. https://doi.org/10.1016/j.catena.2016.07.018

Kutter T, Louwagie G, Schuler J, Zander P, Helming K, Hecker J-M (2011) Policy measures for agricultural soil conservation in the European Union and its member states: policy review and classification. Land Degrad Dev 22:18–31

Lamonica GR, Recchioni MC, Chelli FM, Salvati L (2020) The efficiency of the cross-entropy method when estimating the technical coefficients of input–output tables. Spat Econ Anal 15(1):62–91

Latorre JG, García-Latorre J, Sanchez-Picón A (2001) Dealing with aridity: socio-economic structures and environmental changes in an arid Mediterranean region. Land Use Policy 18. Using and Shaping the Land: 53–64. https://doi.org/10.1016/S0264-8377(00)00045-4

Le Houérou HN (1993) Land degradation in Mediterranean Europe: can agroforestry be a part of the solution? A prospective review. Agrofor Syst 21:43–61. https://doi.org/10.1007/BF00704925

Leontidou L (1990) The Mediterranean City in Transition: Social Change and Urban Development. Cambridge Human Geography. Cambridge University Press, Cambridge. https://doi.org/10.1017/CBO9780511522208

Maes J, Jacobs S (2017) Nature-based solutions for Europe's sustainable development. Conserv Lett 10:121–124. https://doi.org/10.1111/conl.12216

Máñez Costa MA, María EM, Fraser EDG (2011) Socioeconomics, policy, or climate change: what is driving vulnerability in Southern Portugal? Ecol Soc 16. https://doi.org/10.5751/ES-03703-160128

Marathianou M, Kosmas C, St Gerontidis, Detsis V (2000) Land-use evolution and degradation in Lesvos (Greece): a historical approach. Land Degrad Dev 11:63–73. https://doi.org/10.1002/(SICI)1099-145X(200001/02)11:1%3c63::AID-LDR369%3e3.0.CO;2-8

Mavrakis A, Papavasileiou C, Salvati L (2015) Towards (Un)sustainable urban growth? Industrial development, land-use, soil depletion and climate aridity in a Greek agro-forest area. J Arid Environ 121:1–6. https://doi.org/10.1016/j.jaridenv.2015.05.003

Mensching HG (1986) Desertification in Europe? A critical comment with examples from Mediterranean Europe. In: Fantechi R, Margaris NS (eds) Desertification in Europe. Dordrecht: Springer Netherlands, pp 3–8. https://doi.org/10.1007/978-94-009-4648-4_1

Montanarella L (2007) Trends in land degradation in Europe. In: Sivakumar MKV, Ndiang'ui N (eds) Climate and Land Degradation. Environmental Science and Engineering. Springer, Berlin, Heidelberg, pp 83–104. https://doi.org/10.1007/978-3-540-72438-4_5

Moonen AC, Ercoli L, Mariotti M, Masoni A (2002) Climate change in Italy indicated by agrometeorological indices over 122 years. Agric For Meteorol 111:13–27. https://doi.org/10.1016/S0168-1923(02)00012-6

Morelli VG, Rontos K, Salvati L (2014) Between suburbanisation and re-urbanisation: revisiting the urban life cycle in a Mediterranean compact city. Urban Res Pract 7:74–88. https://doi.org/10.1080/17535069.2014.885744

Müller-Nedebock D, Chaplot V (2015) Soil carbon losses by sheet erosion: a potentially critical contribution to the global carbon cycle. Earth Surf Proc Land 40:1803–1813. https://doi.org/10.1002/esp.3758

Munafò M, Marinosci I (2018) Territorio. Processi e trasformazioni in Italia. 296. ISPRA, Roma, Italy

Munafò M, Tombolini I (2014) Rapporto consumo di suolo in Italia. 248. ISPRA, Roma, Italy

Munafò M, Salvati L, Zitti M (2013) Estimating soil sealing rate at national level—Italy as a case study. Ecol Ind 26:137–140. https://doi.org/10.1016/j.ecolind.2012.11.001

Munafò M, Strollo A, Zitti M, Salvati L (2011) Soil sealing e urban sprawl nei territori in transizione: una prospettiva italiana. Rivista Geografica Italiana 118:269–296

Nesshöver C, Assmuth T, Irvine KN, Rusch GM, Waylen KA, Delbaere B, Haase D et al (2017) The science, policy and practice of nature-based solutions: an interdisciplinary perspective. Sci Total Environ 579:1215–1227. https://doi.org/10.1016/j.scitotenv.2016.11.106

Novara A, Gristina L, Guaitoli F, Santoro A, Cerdà A (2013) Managing soil nitrate with cover crops and buffer strips in Sicilian vineyards. Solid Earth 4:255–262. https://doi.org/10.5194/se-4-255-2013

Oral HV, Carvalho P, Gajewska M, Ursino N, Masi F, van Hullebusch ED, Kazak JK et al (2020) A review of nature-based solutions for urban water management in European circular cities: a critical assessment based on case studies and literature. Blue-Green Syst 2:112–136. https://doi.org/10.2166/bgs.2020.932

Orr BJ, Cowie AL, Castillo Sanchez VM, Chasek P, Crossman ND, Erlewein A, ... & Tengberg AE (2017) Scientific conceptual framework for land degradation neutrality. Bonn, Germany: United Nations Convention to Combat Desertification (UNCCD), pp 1–98

Oueslati W, Alvanides S, Garrod G (2015) Determinants of urban sprawl in European cities: Urban Studies. Sage, London UK. https://doi.org/10.1177/0042098015577773

Panagos P, Katsoyiannis A (2019) Soil erosion modelling: the new challenges as the result of policy developments in Europe. Environ Res 172:470–474. https://doi.org/10.1016/j.envres.2019.02.043

Pearlmutter D, Theochari D, Nehls T, Pinho P, Piro P, Korolova A, Papaefthimiou S et al (2020) Enhancing the circular economy with nature-based solutions in the built urban environment: green building materials, systems and sites. Blue-Green Syst 2:46–72. https://doi.org/10.2166/bgs.2019.928

Perrin C, Nougarèdes B, Sini L, Branduini P, Salvati L (2018) Governance changes in peri-urban farmland protection following decentralisation: a comparison between Montpellier (France) and Rome (Italy). Land Use Policy 70:535–546

Pili S, Serra P, Salvati L (2019) Landscape and the city: Agro forest systems, land fragmentation and the ecological network in Rome, Italy. Urban For Urban Green 41:230–237. https://doi.org/10.1016/j.ufug.2019.04.016

Polyzos S, Christopoulou O, Minetos D, Filho WL (2008) An overview of urban-rural land use interactions in Greece. Int J Agric Resour Gov Ecol 7:276. https://doi.org/10.1504/IJARGE.2008.018330

Prasad VK, Badarinth KVS (2004) Land use changes and trends in Human Appropriation of above Ground Net Primary Production (HANPP) in India (1961–98). Geogr J 170:51–63. JSTOR

Puigdefábregas J, Mendizabal T (1998) Perspectives on desertification: Western Mediterranean. J Arid Environ 39:209–224. https://doi.org/10.1006/jare.1998.0401

Rahman, Md. R, Shi ZH, Chongfa C (2009) Soil erosion hazard evaluation—An integrated use of remote sensing, GIS and statistical approaches with biophysical parameters towards management strategies. Ecol Model 220: 1724–1734. https://doi.org/10.1016/j.ecolmodel.2009.04.004

Rapport DJ, Costanza R, McMichael AJ (1998) Assessing ecosystem health. Trends in Ecology & Evolution 13(10):397–402

Recanatesi F, Clemente M, Grigoriadis E, Ranalli F, Zitti M, Salvati L (2016) A fifty-year sustainability assessment of Italian Agro-Forest Districts. Sustainability 8:32. https://doi.org/10.3390/su8010032

Renzi G, Canfora L, Salvati L, Benedetti A (2017) Validation of the soil Biological Fertility Index (BFI) using a multidimensional statistical approach: a country-scale exercise. CATENA 149:294–299. https://doi.org/10.1016/j.catena.2016.10.002

Rodrigo-Comino J (2018) Five decades of soil erosion research in "terroir" The State-of-the-Art. Earth-Sci Rev 179:436–447. https://doi.org/10.1016/j.earscirev.2018.02.014

Rodrigo-Comino J, Brings C, Lassu T, Iserloh T, Senciales JM, Martínez Murillo JF, Ruiz Sinoga JD, Seeger M, Ries JB (2015) Rainfall and human activity impacts on soil losses and rill erosion in vineyards (Ruwer Valley, Germany). Solid Earth 6:823–837. https://doi.org/10.5194/se-6-823-2015

Romano B, Zullo F (2014) Land urbanization in Central Italy: 50 years of evolution. J Land Use Sci 9:143–164. https://doi.org/10.1080/1747423X.2012.754963

Safriel UN (2007) The Assessment of Global Trends in Land Degradation. In: Sivakumar MVK, Ndiang'ui N (eds) Climate and Land Degradation. Environmental Science and Engineering. Springer, Berlin, Heidelberg, pp 1–38. https://doi.org/10.1007/978-3-540-72438-4_1

Salvati L (2014a) A socioeconomic profile of vulnerable land to desertification in Italy. Sci Total Environ 466–467:287–299. https://doi.org/10.1016/j.scitotenv.2013.06.091

Salvati L (2014b) Agro-forest landscape and the 'fringe' city: a multivariate assessment of land-use changes in a sprawling region and implications for planning. Sci Total Environ 490:715–723. https://doi.org/10.1016/j.scitotenv.2014.05.080

Salvati L, Carlucci M (2011) The economic and environmental performances of rural districts in Italy: Are competitiveness and sustainability compatible targets? Ecol Econ 70:2446–2453. https://doi.org/10.1016/j.ecolecon.2011.07.030

Salvati L, Carlucci M (2015) Towards sustainability in agro-forest systems? Grazing intensity, soil degradation and the socioeconomic profile of rural communities in Italy. Ecol Econ 112:1–13. https://doi.org/10.1016/j.ecolecon.2015.02.001

Salvati L, Carlucci M (2016) Patterns of Sprawl: the socioeconomic and territorial profile of dispersed urban areas in Italy. Reg Stud 50:1346–1359. https://doi.org/10.1080/00343404.2015.1009435

Salvati L, Ferrara A (2015) Validation of MEDALUS Fire Risk Index using Forest Fires Statistics through a multivariate approach. Ecol Ind 48:365–369. https://doi.org/10.1016/j.ecolind.2014.08.027

Salvati L, Morelli VG (2014) Unveiling Urban Sprawl in the Mediterranean Region: towards a latent urban transformation? Int J Urban Reg Res 38:1935–1953. https://doi.org/10.1111/1468-2427.12135

Salvati L, Zambon I (2018) Does residential mobility anticipate urban growth? The importance of the local socioeconomic context in a European Metropolitan Region. Popul Rev 57. https://doi.org/10.1353/prv.2018.0005

Salvati L, Zitti M (2009) The environmental "risky" region: identifying land degradation processes through integration of socio-economic and ecological indicators in a multivariate regionalization model. Environ Manage 44:888. https://doi.org/10.1007/s00267-009-9378-5

Salvati L, Zitti M (2011) Economic growth vs. land quality: a multidimensional approach in Italy. J Environ Planning Manage 54:733–748. https://doi.org/10.1080/09640568.2010.528612

Salvati L, Zitti M (2012) Monitoring vegetation and land use quality along the rural–urban gradient in a Mediterranean Region. Appl Geogr 32:896–903. https://doi.org/10.1016/j.apgeog.2011.09.002

Salvati L, Bajocco S, Ceccarelli T, Zitti M, Perini L (2011) Towards a process-based evaluation of land vulnerability to soil degradation in Italy. Ecol Ind 11:1216–1227. https://doi.org/10.1016/j.ecolind.2010.12.024

Salvati L, Carlucci M, Grigoriadis E, Chelli FM (2018a) Uneven dispersion or adaptive polycentrism? Urban expansion, population dynamics and employment growth in an 'ordinary' city. Review of Regional Research 38:1–25. https://doi.org/10.1007/s10037-017-0115-x

Salvati L, Ferrara A, Chelli F (2018b) Long-term growth and metropolitan spatial structures: an analysis of factors influencing urban patch size under different economic cycles. Geogr Tidsskr-Danish J Geogr 118:56–71. https://doi.org/10.1080/00167223.2017.1386582

Salvati L, Gemmiti R, Perini L (2012) Land degradation in Mediterranean urban areas: an unexplored link with planning? Area 44:317–325. https://doi.org/10.1111/j.1475-4762.2012.01083.x

Salvati L, Guandalini A, Carlucci M, Chelli FM (2017a) An empirical assessment of human development through remote sensing: evidences from Italy. Ecol Ind 78:167–172. https://doi.org/10.1016/j.ecolind.2017.03.014

Salvati L, Petitta M, Ceccarelli T, Perini L, Di Battista F, Scarascia MEV (2008) Italy's renewable water resources as estimated on the basis of the monthly water balance. Irrig Drain 57:507–515. https://doi.org/10.1002/ird.380

Salvati L, Sateriano A, Grigoriadis E (2016a) Crisis and the city: profiling urban growth under economic expansion and stagnation. Lett Spat Resour Sci 9:329–342. https://doi.org/10.1007/s12076-015-0160-4

Salvati L, Tombolini I, Gemmiti R, Carlucci M, Bajocco S, Perini L, Ferrara A, Colantoni A (2017b) Complexity in action: untangling latent relationships between land quality, economic structures and socio-spatial patterns in Italy. PLoS ONE 12:e0177853. https://doi.org/10.1371/journal.pone.0177853

Salvati L, Tombolini I, Perini L, Ferrara A (2013) Landscape changes and environmental quality: the evolution of land vulnerability and potential resilience to degradation in Italy. Reg Environ Change 13:1223–1233. https://doi.org/10.1007/s10113-013-0437-3

Salvati L, Zitti M, Ceccarelli T (2007) Integrating economic and environmental indicators in the assessment of desertification risk: a case study. Appl Ecol Environ Res 6:129–138. https://doi.org/10.15666/aeer/0601_129138

Salvati L, Zitti M, Perini L (2016b) Fifty years on: long-term patterns of land sensitivity to desertification in Italy. Land Degrad Dev 27:97–107. https://doi.org/10.1002/ldr.2226

Scalenghe R, Marsan FA (2009) The anthropogenic sealing of soils in urban areas. Landscape Urban Plan 90:1–10. https://doi.org/10.1016/j.landurbplan.2008.10.011

Shalaby A, Tateishi R (2007) Remote sensing and GIS for mapping and monitoring land cover and land-use changes in the Northwestern coastal zone of Egypt. Appl Geogr 27:28–41. https://doi.org/10.1016/j.apgeog.2006.09.004

Singer MJ, Le Bissonais Y (1998) Importance of surface sealing in the erosion of some soils from a Mediterranean Climate. Geomorphology 28:79–85

Sivakumar MVK (2007) Interactions between climate and desertification. Agric For Meteorol 142:143–155. https://doi.org/10.1016/j.agrformet.2006.03.025

Smiraglia D, Ceccarelli T, Bajocco S, Salvati L, Perini L (2016) Linking trajectories of land change, land degradation processes and ecosystem services. Environ Res 147:590–600. https://doi.org/10.1016/j.envres.2015.11.030

Stavi I, Lal R (2015) Achieving zero net land degradation: challenges and opportunities. J Arid Environ 112:44–51. https://doi.org/10.1016/j.jaridenv.2014.01.016

Symeonakis E, Karathanasis N, Koukoulas S, Panagopoulos G (2016) Monitoring sensitivity to land degradation and desertification with the environmentally sensitive area index: the case of Lesvos Island. Land Degrad Dev 27:1562–1573. https://doi.org/10.1002/ldr.2285

Tanrivermis H (2003) Agricultural land use change and sustainable use of land resources in the Mediterranean Region of Turkey. J Arid Environ 54:553–564. https://doi.org/10.1006/jare.2002.1078

Terzi F, Bolen F (2009) Urban Sprawl measurement of Istanbul. Eur Plan Stud 17:1559–1570. https://doi.org/10.1080/09654310903141797

Tombolini I, Zambon I, Ippolito A, Grigoriadis S, Serra P, Salvati L (2015) Revisiting "Southern" Sprawl: urban growth, socio-spatial structure and the influence of local economic contexts. Economies 3:237–259. https://doi.org/10.3390/economies3040237

UNCCD (2015) *Report of the Conference of the Parties on its twelfth session.* ICCD/COP(12)/20/Add.1. Paris, France

United Nations (2011) Cities and climate change: global report on human settlements. 001/11E, 003/11E. UN-HABITAT, London, UK

(USA) Dodds, Felix (University of North Carolina & Tellus Institute, Ambassador David (Permanent Mission of Ireland to the United Nations USA) Donoghue, and Jimena (International Peace Institute USA) Leiva Roesch (2016) Negotiating the Sustainable Development Goals: A transformational agenda for an insecure world. Illustrated Auflage. Taylor & Francis Ltd, London and New York

Vanwalleghem T, Amate JI, González M, de Molina D, Fernández S, Gómez JA (2011) Quantifying the effect of historical soil management on soil erosion rates in Mediterranean olive orchards. Agr Ecosyst Environ 142:341–351. https://doi.org/10.1016/j.agee.2011.06.003

Warkentin BP (1995) The changing concepts of soil quality. J Soil Water Conserv 50:226–228

Zambon I, Benedetti A, Ferrara C, Salvati L (2018) Soil matters? A multivariate analysis of socioeconomic constraints to urban expansion in Mediterranean Europe. Ecol Econ 146:173–183. https://doi.org/10.1016/j.ecolecon.2017.10.015

Zambon I, Colantoni A, Carlucci M, Morrow N, Sateriano A, Salvati L (2017a) Land quality, sustainable development and environmental degradation in agricultural districts: a computational approach based on entropy indexes. Environ Impact Assess Rev 64:37–46. https://doi.org/10.1016/j.eiar.2017.01.003

Zambon I, Serra P, Sauri D, Carlucci M, Salvati L (2017b) Beyond the 'Mediterranean city': socioeconomic disparities and urban sprawl in three Southern European cities. Geografiska Annaler: Series B, Human Geography 99:319–337. https://doi.org/10.1080/04353684.2017.1294857

Zell A, Mache N, Hübner R, Mamier G, Vogt M, Schmalzl M, Herrmann K-U (1994) SNNS (Stuttgart Neural Network Simulator). In: Skrzypek J (ed) Neural Network Simulation Environments. The Kluwer International Series in Engineering and Computer Science. Springer US, Boston, MA, pp 165–186. https://doi.org/10.1007/978-1-4615-2736-7_9

Chapter 3
Mediterranean Europe, a Fragile Landscape: Metropolitan Growth and Urban Sprawl

Abstract In addition to having a high degree of freedom and self-organization, the Mediterranean city has been outlined as the place where a high degree of spatial, cultural, but also institutional disorder is achieved. The city is sometimes read in its many components as represented by a "difficult order to understand". Therefore, in this chapter, we try to define and describe the main aspects and issues of this fragile landscape. We will discuss the difference between the formation of the metropolis and the settlement disorder, paying attention to some examples from Easter Mediterranean, Italy, Spain, Southern France or Greece. Moreover, the socio-economic structure of Barcelona, Rome and Athens will be used as an example to explain the metropolitan growth and urban sprawl. They appear very different, although they are all located in the Mediterranean basin. Each one has a unique and different territorial configuration and the phenomenon of urban sprawl has adapted differently, following the economic and social connotations of the countries under investigation.

Keywords Urban sprawl · Peri-urban areas · Metropolis · Space

3.1 The Intrinsic Fragility of Semi-Natural Landscapes in the Mediterranean Basin

Coastal and lowland areas in the Mediterranean basin are generally described as vulnerable to land degradation, due to both anthropogenic factors and the impact of climate change. Considering the climate a highly complex system, it can be understood how the terms change, variability, anomaly and balance must concern not single and limited territorial areas or individual atmospheric physical quantities but the global involvement of each variable and all cause-effect interrelationships existing between them. For example, in the area considered, soil degradation certainly derives from the combination of climatic aspects with those related to land management, including the improper use of land that generates spatial disparities in environmental and anthropogenic factors (Gisladottir and Stocking 2005), widening the ecological gap between peri-urban and rural areas (Portnov and Safriel 2004).

I. Tombolini et al., *Land Quality and Sustainable Urban Forms*, Springer Geography,
https://doi.org/10.1007/978-3-030-94732-3_3

The vulnerability to land degradation in the Mediterranean region is therefore the result of the multifaceted and spatially variable interaction between the environmental sphere and the socio-economic system (Montanarella 2007). Although the environmental characteristics of vulnerable areas are similar to those of already degraded lands, some factors such as vegetation, agriculture and policy strategies can mitigate the evolution of land degradation, even in the short term (Biasi et al. 2015a; Rodrigo-Comino et al. 2020). Furthermore, traditional agricultural systems contribute to the eco-compatible uses of rural areas, preserving the quality of the soil and ensuring the functioning of ecosystems in the long term (Siciliano 2008). In Mediterranean Europe, it is common to find economically disadvantaged and marginal rural contexts, moreover in arid environmental conditions (Salvati and Carlucci 2011). In those contexts, which are characterized by a long history of human settlement and intensive land use management (Blondel 2006; Luca Salvati 2014b; Hernández et al. 2015), socio-economic factors and variable biophysical conditions are being influenced by local socio-ecological systems, showing elicit complex responses related to the degradation of natural resources (e.g. Berkes et al. 2000; Kurttila 2001; Kelly et al. 2015). Under these conditions, land degradation, which in the Mediterranean area is intrinsically linked to overgrazing, fires, unsustainable exploitation of water and soil resources, and pollution caused by pesticides and herbicides, proves to be a worrying aspect especially in following the abandonment of lands, soil erosion, rural poverty and the loss of land value (Santos and Cabral 2004; Bajocco et al. 2011; Biasi et al. 2015b). The expansion of degraded areas increasingly involves traditional agricultural systems (Salvati and Zitti 2005; Bajocco et al. 2012; Arnaiz-Schmitz et al. 2018), resulting in a progressive depletion of fertile lands, the loss of biological and economic productivity, soil erosion, habitat fragmentation and the reduction of ecosystem services (Salvati et al. 2007). Among other things, the Mediterranean basin is a geographical region with a significant biodiversity reserve that is at risk for anthropogenic causes (it is a biodiversity hotspot; Myers et al. 2000), which further reinforces the need to carefully monitor its territory and its critical factors (Shaw et al. 2020).

The decrease in high-quality (fertile) agricultural land can also be attributed to urban sprawl, especially in flat and accessible rural districts (Recatalá et al. 2000). Population growth in urban areas, in turn, stimulated an increase in food demand which in some cases led to the intensification of crops (Gardi et al. 2015), which exacerbates land degradation (Kangalawe and Lyimo 2010; Bakr et al. 2012; Kosmas et al. 2016). Especially, in the most marginal areas, characterized by fragile ecosystems, even the slightest environmental conditions are unable to withstand the demographic pressure and support the relative use of the territory (Salvati et al. 2016a).

Ferrara et al. (2014) hypothesized that the recent increase in the level of soil vulnerability is distributed heterogeneously over time and space in the Mediterranean region, affecting areas that have a specific composition of land use and specific socio-economic characteristics, especially for this, which concerns the agro-forest ecosystems surrounding the urban agglomerations and in particular the *shrinking cities*[1] (Richardson 2014). Considering the Eastern Mediterranean countries such as Slovenia in comparison to western European countries, which after the Second World War experienced intense urban growth. The most extended urban system was characterized by a specific polycentric development based on diverse management plans designed by policymakers (C. Di Feliciantonio and Salvati 2015). Most of the plans focused on the reduction of the agricultural sector but it was not followed by intense migration to urban areas and could be better described as moderate growth of urban population (Uršič 2012). After 1991, coinciding with the novelty of free-market conditions and avoiding the socio-economic restrictions, urban sprawl drastically occurred. This was also benefited by the improvements of the highway network and intensive automobilization, with direct implications on more affluent of residents and businesses attraction, consumption of green and suburban areas and change in the role of city centers (Munoz 2003). For example, in Croatia, until 1960, the most populated settlements were the rural areas. However, from the 1960s, due to the intensive industrialization, rural exodus and the development of service industry, demographic changes occurred parallel to the occupation of military barracks (Graovac Matassi 2014). During the last two decades, simultaneously with demographic and economic changes, this situation drastically changed thanks to other factors such as the development of the coastal tourism, carefully rebuilding and upgrading the hotel capacities (Kranjčević and Hajdinjak 2019). For Serbia, Slaev et al. (2018) identified two different groups of factors to identify urban development (urban sprawl and suburbanization) in Serbia: (i) typical such as the higher rates of car ownership and incomes of certain social strata; and, (ii) atypical factors such as cultural traditions and housing preferences. Maksin-Mićić (2008) estimated that the urban population in Serbia increased from 22.5% in 1953 to 56.4% in 2002. This demonstrated that Serbia was less urbanized at the beginning of the century than Bulgaria, Hungary or Romania, but more than Albania and Bosnia and Herzegovina, with similar levels as Croatia or Republic of North Macedonia.

Taking the Italian landscape as an example, a careful observer can observe the countless signs that economic development has left on the territory (Salvati and Zitti 2008, 2009). The landscape that, in the north as in the south, began to lose the characteristics of the widespread rurality that had characterized it in the first half of the last century (Sereni 1996), starting from the second post-war period, to transform itself into progressively urbanized and often highly infrastructural contexts. The phenomena of urban growth and coastalization affect the large urban areas of

[1] Shrinking cities are understood here as cities that have undergone a significant drop in population, emigrated to more peripheral areas. Haase et al. (2013) emphasize the diversity of shrinking cities, interpreting shrinkage as a context where macro-processes produce very different outcomes at the local level, which can only be read by taking into consideration the specific history of the city where they occur. To remain within the Italian landscape on which a focus is placed in this paragraph, the main features of urbanization are not only connected to a long history of political fragmentation of the soil, but also to very different rural models present in the North, Center and South. "It is not only the legacy of large land properties cultivated by wage laborers that makes the difference between North and South, but also the different way in which industrialization formed the entrance into modernity. As already pointed out by Lila Leontidou for the Greek case (1990), it was not the factories that caused the urbanization but the dissolution of the countryside" (Bini et al. 2014).

the Po valley and the agglomerations—seamlessly—of Rome and Naples, as well as the Apulian and Sicilian coasts, the Tuscan-Lazio Maremma, the entire Adriatic coast from Romagna up to Molise. In the same years, intensive agriculture of monoculture and industrial animal husbandry alternated with the high-density industrial settlement, manipulating the typical features of the landscape, reducing that quality which is one of the most promising capital endowments for future growth possibilities (Trisorio 2005; Luca Salvati et al. 2008). A picture of this magnitude is made even more complex by the drastic reduction in public investment.

In Spain or the Southern coast of France, the psychological effects of a growth sustained from above disappear (Hubacek and van den Bergh 2006) which, instead, has seen the territory as a protagonist in the past, with particular reference to those territories, marginal and, at the same time, high landscape vocation, which for years, even during the economic boom, have retained typical features of rurality and biodiversity, currently compromised by a more slender recent development, but also more widespread and spatially widespread (Salvati et al. 2016a). The rural landscape alternates, in fact, urban areas of high production intensity and tourism activities with vast agricultural areas bordering on abandonment (Cuadrado-Ciuraneta et al. 2017). Spanish and French agricultural systems have assessed, from a systemic perspective, the relevance of the various economic, social and environmental interacting dimensions (Díaz-Pacheco and García-Palomares 2014). As main themes, the first dimension identifies the efficient use of resources and the vitality of the agricultural sector, as well as the contribution of the primary sector to the conservation of rural areas (Pirotte and Madre 2011; Morollón et al. 2016; Luca Salvati 2016a). The social dimension refers specifically to human capital and its characteristics on a local scale (De Rosa and Salvati 2016; Mosconi et al. 2020). Finally, the environmental dimension highlights the management and conservation mechanisms of natural resources, in terms of landscape, water resources and soil, demonstrating how only an integrated and systemic assessment, but at the same time synthetic, can provide a correct interpretation of the trajectories observed, with all related policy implications (Quaranta and Salvia 2005).

3.2 From Dispersed Cities to Metropolitan Networks

Understood as the result of a process of co-evolution between humans, nature and culture, reflected in the complexity of the socio-economic organization of the urban space and expressed by the distribution of the different types of settlement. The Mediterranean city (e.g. Figure 3.1) can be considered a product in its own right, belonging to a culture that finds its founding aspects in the particularism and localisms that distinguish it.

Among the typical features of the Mediterranean city, we can count the compactness of the settlement (Fig. 3.2). The strong identity linked to the presence of a dominant historical center, the distinctive and not disjointed features of the surrounding

Fig. 3.1 Lisbon and Porto cityscapes, cities located in important strategic points, at the mouth of the Tagus and Douro Rivers and the foot of the Atlantic Ocean (Photo by the authors [April 2018 and January 2012])

Fig. 3.2 The compact and dense morphology of Mediterranean cities: three examples in comparison (Barcelona on the left, Naples in the center, Athens on the right) (Images were taken from the monograph "Exploring urban complexity" [Tombolini et al. 2017])

countryside. Although often overshadowed from the central city, the development of local communities that evolve according to original lifestyles and consumption.

Although there is no truly unifying framework in which it is possible to frame it, the Mediterranean city, understood as an element of distinction from the development models of the most developed urban systems has been placed at the center of the debate on southern urbanization processes by authors such as Muscarà (1978) and Leontidou (1990). This can be only understood with the awareness that only a comparative analysis of the different urban trajectories can form a territorial synthesis (Chris Couch et al. 2008b), in a context characterized by strong spatial heterogeneities and socio-economic specificities that cannot be assimilated to common paths (Zambon et al. 2017).

For the explanation of the phenomena of southern urban systems, Leontidou (1993) places the need for "local" theories at the center of the debate, rather than general theory. The construction (or reconstruction) of interpretative models aimed at declining the complexity of horizontal and vertical interrelationships within urban society, highlighting the symbolic centrality of the Mediterranean city, the historiographic approach to urban growth, as well as a reflection of the qualifying points of this interpretative strand, are the (missing) roles of urban and regional planning (Di Feliciantonio et al. 2018). Among the elements on which it rotates there are certainly: (i) a peculiar morphology, with a compact or semi-compact urban landscape, with narrow road axes, with a tortuous and irregular design that sometimes follows the relief, which makes the periphery physically interconnected with the center; and, (ii) a mixed and entropic use of building land, as opposed to the more regular zoning typical of the northern city. These peculiarities have an effect on urban culture, on the polarity of the city center, on the poor effectiveness of planning at various spatial scales, on the widespread use of the informal economy, but also cosmopolitanism, probably one of the most recognized aspects of all large Mediterranean cities (Salvati et al. 2018b).

The particularistic elements mentioned above allow us to reflect on the Mediterranean cities, classifying them not as a homogeneous whole with a strong regional characterization but: (i) as a complex of historical and traditional urban areas; (ii) functionally heterogeneous and spatially fragmented; (iii) with socio-economic profiles deriving from different and sometimes conflicting development paths, which cannot be traced back to a common interpretative matrix. Cities are objectively very different from each other, daughters of a region with undoubtedly similar environmental, historical, architectural elements, but which continues to outline forms of development articulated and changeable, between "globalistic" thrusts, "regional" suggestions and "local" spontaneous (Pace 1996). The result of these thrusts, reflected in the settlement and social stratification of the past, is grafted onto the most recent urban dynamics in a fragmented transformation matrix and without uniqueness (Morelli and Salvati 2010; Morelli et al. 2014). The essence of that "slow transition" that Lila Leontidou described for Athens in 1990 and which can be generalized twenty years later, albeit, with appropriate caution, too many other urban contexts of the Mediterranean in continuous evolution should be particularly understood. New urban plots, socio-demographic dynamics, economic structures, but also so strongly linked to widely settled models (Salvati 2014a).

While the traditional, compact and dense historic cities, developed along radial axes that branch out regularly from the center and functionally focused on the manufacturing industry and the tertiary sector, appeared as an orderly and understandable organ (Ciommi et al. 2019). The process of branch expansion and the formation of areas increasingly large and less compact metros is the result of sedimentation of additions, made in different times and ways, which have affected the territory, modulating it according to an unstable reading frame. Since the 1970s, the Mediterranean area has been hit by urban sprawl, an intriguing (and intricate) socio-economic phenomenon, as a result of which cities have taken on a polarized spatial structure that sometimes emphasizes disparities between the population (Ciommi et al. 2018; Salvati et al. 2018a, 2018b). Many of the reference economic centers have undergone a progressive fragmentation, assuming a spatial organization difficult to classify as "polycentric" and with significant infrastructure barriers (Carlucci et al. 2018).

At this point, it is legitimate to ask ourselves which paths the new urban areas of Southern Europe have recently undertaken and what future awaits them in the short term. For example, it is probably useful to ask about the "global" and "systemic" meaning of important convergence processes. This started relatively recently, toward more developed and performing urban models (at least in economic terms) in the relationships between cities and territories (Amin and Thrift 2005), as well as on the future developments that arise from them, possibly separating what can be considered as specific and limited to specific themes, from a broader process of sedimentation of urban characters (Salvati et al. 2019).

The most evident change taking place is the explosion of the city: the concentrated city has started to lose population, activities and services which have found a more opportune and convenient location in the large territory, outside the founding walls, while inside they are establishing a process of social, professional and activity polarization (Salvati and Morelli 2014). Indovina (2009) argues that the formation of the widespread city is essentially the outcome of a need for the city, that is, the relationships (economic, social, etc.) that are constitutive of the city. This phenomenon does not correspond to a refusal toward the so-called "traditional" city. This rather expresses a need for a "different" city, both by those who had not previously had direct experience of urban life (the inhabitants of the countryside) and by those who came from a city assumed to be unsatisfactory and therefore looking for a new and better urban condition. The realization of social, economic, cultural, friendly, urban-type relationships in a morphologically non-urban context clarifies what is the founding datum of the urban condition itself: not the walls, not a given morphological form, but the constituted and constituting society. The widespread city, therefore, functions like a concentrated city without having its density characteristics (Chelli et al. 2016).

From urbanization, there is, in general, extensive use of the territory and intensive use of the city. It is a traditional phenomenon but which today is characterized differently, both for the continuity in some areas of the urbanized territory and for an increased density of the city (Salvati 2018).

More and more, in many areas the territory is no longer characterized by being a countryside from which cities emerge, but rather by being a huge city with enclosed countryside areas (Figs. 3.3 and 3.4).

In terms of urban planning and services, the meeting, and sometimes confrontation, of the policy of local authorities with the "demand" coming from a growing and changing population in social and "experience" terms has determined an increase in the supply of infrastructures and collective services. The phenomenon has been homogeneous neither from the spatial point of view nor from the sectoral one: a mix of concentration and diffusion seems to be the recurring typology, but with the sole objective of feeding a supply system for the diffused city, for a population that is, located in a large and low-intensity territory (Ciommi et al. 2017b). The evolution of this type of settlement is due to the possibilities offered by ever new technologies; to changes in lifestyle; looking for possible new economic opportunities; the split of the urban condition from the urban morphology and, in summary, the possibility of benefiting from the advantages of the agglomeration (occasions, information, sociability, etc.) without agglomerating.

The consolidation of the widespread city, including the densification processes resulting from spontaneity or timid public interventions, the need to escape the urban cost of the concentrated city and a different location of users and customers have led to changes in the location, probably improving it, both of government centers

Fig. 3.3 The landscape of the three Spanish coastal cities and its diffusion (Photo by the authors from Cádiz [above-left], Málaga [above-right], Almería [below-left] and Palma de Mallorca [below-right])

and the same poles of excellence.[2] Some of these follow the trend of diffusion and are located in the wide territory, creating new opportunities and above all new territorial organizations (Gigliarano and Chelli 2016). In addition, the consistency of the population settled in the vast area, which continues to show increasing (one could say of the size of a metropolis), leads to the emergence of metropolitan-type private and public services (Chelli and Rosti 2002). The difference between cities and metropolises is therefore not only quantitative but also qualitative: the size of the population allows the activation of "rare" services that satisfy a very large user base.

To conclude, the transition from the widespread city to the territorial metropolis constitutes a stimulus to reflect on the transformations of the territorial organization in southern Europe concerning other areas of the world. It has been seen that, regardless of the specific organization of the individual realities, a process of "metropolization" of the territory is also taking place in the Mediterranean area

[2] The fact that the major centers tend to lose the highest-level services and the poles of excellence on the one hand reduces the hierarchy of the territory, on the other it creates a metropolitan structure, not concentrated. This phenomenon is seen with concern by individual local administrations, which exercise their attention on the point and who perhaps are not yet ready for a management of the large area.

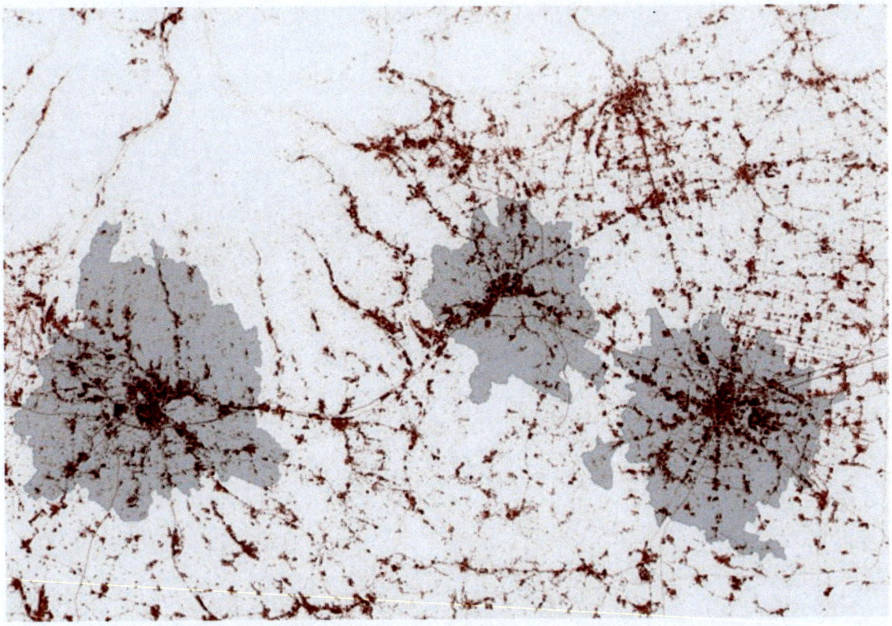

Fig. 3.4 A very high-resolution map showing the waterproofed soil, as a reflection of the widespread city, in the Veneto area (from west to east Verona, Vicenza, Padua)

(Salvati and Lamonica 2020). It shows the often self-organized realization of a specific metropolis, not concentrated, which uses at most the resources of the territory. This is a phenomenology that presents itself according to various degrees of development and maturation. The new territorial structure, containing more or less ample continuity solutions, has open spaces that can be used in part in agricultural production and in part allow better organization and networking of parks, woods and equipped areas. More generally, these represent an opportunity to rethink a new role, socio-environmental rather than productive, for the entire metropolitan area (Ciommi et al. 2017a). This allows managing the infrastructure bet as a tool for sustainable development, placing the protection of green areas and the requalification of the building heritage as a priority on the political agenda and, above all, reconsidering the role of the urban region in a wider process of polycentric development on a national scale (Delladetsima 2006).

3.3 The Mediterranean City as an Entropic and Disorder Space

In addition to having a high degree of freedom and self-organization, the Mediterranean city has been outlined as the place where a high degree of spatial, cultural, but also institutional disorder is achieved. The city is sometimes read in its many

Fig. 3.5 The settlement disorder of the Mediterranean city represented in the low-density forms of the periurban of Granada (Photographies taken by the authors)

components as represented by a "difficult order to understand" (Minca 2003). In this perspective, metropolitan sprawl seems to have contributed to determining a greater level of functional entropy and settlement disorder with the mixture of land uses sometimes in conflict with each other, far from that mix of uses typical of peri-urban mosaic, still in agricultural part, still partly forest (Salvati et al. 2012a) (Gargiulo Morelli and Salvati 2010). Without narrative rules, and with few planning certainties, the Mediterranean metropolis is "[…] the result of a multitude of choices, which are all rational or aspire to be so, but which obey different logics, in antagonism with each other" (Pace 1996).

The settlement disorder is reflected not only in the geographical and socio-spatial reading of the system but also the management of the existing and in the planning of the near future (Couch et al. 2008a). Urban forms, infrastructures, public places, landmarks, although resulting from a multitude of choices, the majority reasonably effective in themselves, are not organized in a unitary way and seem to obey fragmented and deeply contrasting logics (Pozzolo 2001), in a disunited fabric and antagonism to the diffusion without the logic of the remote periurban (Munoz 2003). In this matrix, planning sometimes manifests an objective difficulty of interpretation and unity of visions (Salvati et al. 2012b). Again, according to Pace (1996): "[…], local authorities should intervene to dictate the 'rules of the game': but we know they cannot do it, because the culture of the 'metropolis' itself, once it has been dismissed as a 'Bolshevik', is a libertarian, aspires to deregulation, is in short, open to the initiative of the large multinational as of the small abusive builder".

A source of wide debate is how features of building spontaneity have been integrated into (and declined by) local traditions, regional economic structures and national regulations (Giannakourou 2005). What is certain is that this process has made the urban patterns even more heterogeneous, the sprawl even more aggressive, the fragmentation of the built fabric and the peri-urban landscapes even more marked (Salvati 2012).

The most recent urban spread, albeit partially planned in most southern urban contexts (Morelli and Salvati 2010), appears as a further element of disorder in the system (Fig. 3.7).

The large southern urban regions appear, in their growth not yet completely dormant, as dynamic and articulated organisms, the complexity of which is the result of previous interventions, generally punctual, rarely of the system, whose disorder is the result of yesterday's lack of choices and today, which have weighed heavily on the spatial organization of the territories (Colantoni et al. 2015). By integrating form and functions, following this "additive" model, it is difficult to read the urban plot and interpret the multiple relationships with the surrounding areas. The spatial disorder remains a typical feature of the Mediterranean city, perhaps enhanced by the recent suburbanization, which has consolidated its territorial disparities and social inequalities (Colantoni et al. 2016).

3.4 Three Protagonists of Urban Sprawl

The millennial history that affects the Mediterranean basin has always generated a certain charm, becoming the subject of various studies and insights. In particular, the northern area, located on the European continent, has aroused great interest from scholars. It is precisely because of their long past that European cities bordering the Mediterranean Sea are characterized by social, political, cultural and economic factors. Although, as discussed in the previous paragraph, these cities resemble each other in some respects, they also have differences that should not be underestimated. This mix of equalities, which derive from a territorial area that is affected by the same climatic and historical components, and inequalities, due to other factors that have affected differentially, makes each city particular and unique (Leontidou 1990).

Here, we want to focus on three cities as example considered protagonists of the dispersed urbanization: Barcelona, Rome and Athens. All three are Mediterranean cities that play an important role in their national sphere but also in the European and international spheres. -Barcelona—an important Spanish city overlooking the Mediterranean Sea, at the center of international interest for multiple economic, cultural and tourist aspects—Rome—the capital of Italy, with a very rich history and heritage, appreciated and recognized worldwide—Athens—equally interesting from a historical and cultural point of view, a Greek capital and an important port area—have numerous points of similarity that do not concern only the geographical and climatic conditions (Zambon et al. 2018; Polinesi et al. 2020).

They have undergone a rapid transition from their traditional compact settlement model to a more dispersed form, characterized by an impressive sprawl around their urban area (Schneider and Woodcock 2008). Urban transformations, driven by demographic change, socio-economic processes, institutional and cultural factors together with those of a territorial and environmental nature, no longer concern only the urban nuclei but also, and above all, the vast areas that surround them. For this reason, as it is no longer possible to analyze cities by confining issues based on their administrative limits, it is necessary to consider them in a wider system, defined as a "metropolitan area" (Salvati et al. 2018c).

The sprawl gave rise to a landscape characterized by suburban areas with low-density residential settlements, often lacking architectural quality, which gradually replaces the traditional agricultural and forest areas, thickening along the main roads (Catalán et al. 2008; Arapoglou and Sayas 2009; Cecchini et al. 2019). The new inhabitants, protagonists of this suburbanization process, generate a new lifestyle and new flows of exchange with the city; in addition to the desire to own a single house surrounded by greenery, they are also heavily dependent on their individual and private transport (Torrens 2008). The comparison between the three metropolitan areas probably reveals how the sprawl occurred in different ways, influencing differently both the pattern of the settlement morphology and some processes related to the spatial distribution of economic functions, social and territorial disparities, and political factors and cultural structures consolidating the role of peri-urban areas (Zambon et al. 2019).

3.4.1 Settlement and Morphological Aspects

As previously announced, the three metropolises considered have undergone a transition from a compact urban model toward a more widespread configuration, which has involved the current suburban context. They are also united by a similar deregulated past, characterized by non-authorism and informality, due to an urban design without a precise spatial orientation and efficient planning (Costa et al. 1991; Gibelli and Salzano 2006). Barcelona, Rome and Athens have the typical socio-spatial configuration of the compact cities of Southern Europe as the structure are fragmented and polarized, underlining the differences, both of the new settlements and those already consolidated, evident also on a regional scale (Delladetsima 2006; Maloutas 2007; Chris Couch et al. 2008b). Finally, they are cities that play a capital role and that have hosted competitive events of global importance, such as the Olympic Games in the case of Athens and Barcelona, during which urban transformation mainly concerned the metropolitan area. Barcelona, Athens and Rome differ in turn by a variety of factors of various nature, interacting with each other and are linked to the history of their settlement dynamics (Salvati and De Rosa 2014).

Fig. 3.6 The compact and dense morphology of Barcelona (The photo was taken by the architect Serena D'Amora [February 2018])

Since 1950, new neighborhoods have sprung up in Barcelona, ready to be the residence of that population that migrated from rural areas to the city: a phenomenon very similar to what happened in Italy. This phase saw the formation of urban suburbs without open spaces, social services, infrastructure and public transport. At the same time, there was also the phenomenon of second homes, which arose not only in coastal locations but also in the central area of the metropolitan area of Barcelona (Fig. 3.6), usually on cheap and illegally land (Catalán et al. 2008).

In those years, there was a dispersion of economic activities and the population in the urban area and the municipalities of the first belt, thus saturating the whole part of the "Eixample". The discontinuous growth has led to the formation of islands or urban development areas in the marginal areas of the metropolitan area. This phenomenon of segregation and exclusion, not only spatial but also social, is associated with las "baraccas", areas of physical and social degradation in large areas of the historic center, close to industrial areas (Fig. 3.7). Here, those neighborhoods were also born at low-cost thanks to public building initiatives. It was at that moment that the first phenomenon of suburbanization started (Mazzoleni 2009).

In the 1950s, residential dispersal and the process of industrial decentralization had already crossed the borders of the municipality of Barcelona, although it is necessary to wait until the 1960s to identify a significant trend toward suburbanization (Dura-Guimera 2003).

In the 1960s and 1970s, a large number of people living in Barcelona bought houses in the rural–urban fringe, often in hilly areas, making the population of the

Fig. 3.7 Las baraccas in the El Carmelo neighborhood in Barcelona (The photo was taken from the site https://www.pinterest.it/pin/335377503477652427/)

metropolitan area grow from 2.5 to 4 million people (Sauri 2003), growth which was also associated with speculation and abusive phenomena. The result of speculative pressure reached its peak in the 60s and 70s with the so-called "Spanish miracle". Since 1975, the migratory flows of the population have balanced even if the residential dispersion has continued to grow. Between 1975 and 1986, exactly with the Spanish economic downturn, urban growth remained rather low (Catalán et al. 2008). For the first time in centuries, the city began to lose inhabitants and the tendencies toward urban decentralization and urban expansion became more acute. Another powerful factor was the increase in tertiary services, which led to the migration of the population from the city to the urban region (Paül and Tonts 2005).

Barcelona has been able to reap new opportunities for revitalization with the restoration of democracy, independent institutions (1978–1980) and the accession of the nation to the European Community in 1986 (Dura-Guimera 2003). As happened in Athens, Barcelona was the host city of the Olympic Games in 1992. The local government (social democratic) not only promoted the Games as a modernization project but also tried to redistribute the offers of residential accommodation throughout the area underground (McNeill 1999). Barcelona currently represents a meeting place for cultures and people from very different and migratory flows represent an important social phenomenon, the basis of urban deconcentration and sprawl (Lesthaeghe and Lopez-Gay 2013).

The recent history and urban development of Rome have some similarities to the Barcelona case. Before going into urban diffusion, the Roman territory saw a period of "concentration", from the post-war period to the 1970s, characterized by a high unemployment rate, a low activity rate, economic structures based on trade and settlements having a monocentric spatial organization (Seronde-Babonaux 1983; Costa et al. 1991; Krumholz 1992; Insolera 2001). Urban growth was fueled by massive immigration from southern Italy, saturating urban gaps and invading other portions of the territory that are still underdeveloped. In addition, as mentioned above, the history of Rome was marked by the phenomenon of illegal conduct, which explains why urban sprawl has manifested itself without rules. It is precisely in the 1960s, with the economic boom, that in Rome there is strong building anarchy, an infrastructure of the territory not in line with urban planning rules (Salvati and Carlucci 2016).

This is how Italo Insolera (architect, urban planner, Italian historian) describes the cynical indifference toward the fate of the city in that decade, in chapter XXIV of Roma Moderna: "In the 1950s the owners of the city knew that their every move was immediately registered: communists, socialists, left-wing Catholics, National Urban Planning Institute, Italia Nostra, information press, dozens of men of culture and influential personalities were ready to go down in the field, to sign appeals and petitions, to take part in debates, conferences. In the 1960s, the owners of the city knew that their moves would hardly and rarely find space in the news".

It is above all the gigantic and inhuman suburbs that had been built in the years of fascism and in the first decade of the post-war period to interest Insolera, whose shots of the suburbs rarely portray people, but more and more cement, asphalt and traffic that seem to absorb every space (Fig. 3.8). As argued by Martinotti (1993), the 1970s marked the end of the physical and demographic growth of large Italian cities, giving rise to a more widespread growth at the metropolitan level. The literature describes the city of Rome in four characteristics: the historical city; the speculative one, built in the 50s and 70s thanks to strong private companies; the public one; and the abusive one (Della Seta and Della Seta 1988). Since 1980, urban sprawl has led to the rapid spread of medium-density settlements, while social disparities and economic structure have remained relatively unchanged (Clemente et al. 2018).

The urban development of Rome took place along the main radial axes creating a "hybrid" landscape, formed by fragmented areas and by the overlap of widespread settlements and rural areas. In the early 1990s, like many European cities, Rome faced the problems of economic restructuring and global competition, feeling the need to be a freer city in international trade, attractive for the flows of capital and people (Forino et al. 2015).

The economy is based on activities typical of global cities: there is a strong presence of services, the construction industry and new technologies (such as electronics, telecommunications and biotechnology). It is also a high tourist city. The attraction that generates the capital causes a strong tourist interest that feeds regional disparities along the urban–rural gradient, stimulating new economic opportunities only in the urban context (Munafò et al. 2013). Among the various cities of southern Europe, Rome has some shortcomings in terms of population growth, employment,

Fig. 3.8 Residential complex located on the edge of the park of Villa Gordiani (Rome), 1959 (Photo by Italo Insolera)

the economy, especially concerning the tertiary sector, social integration. This is due in part to strong urban congestion, a considerable concentration of population in certain areas and economic polarization which are opposed to the polycentric urban form, which is more competitive, thanks to its expanding production. It is a densely populated city if we think that half of the inhabitants of the Lazio region live in the capital, and is characterized by an extremely fragmented suburban landscape, in which the marginal territories have distinct social and economic characters, and are highly dependent on the urban core (Allegretti and Cellamare 2009).

Although there are some similarities with the Spanish and Italian cities, in Athens, which in modern times has failed to take off economically, there has been no phase of particular de-industrialization in the post-war period. The expansion phases of the Greek city have been well-characterized by demographic, economic and social factors; Athens has grown over time in an unplanned way, in some cases through a design consisting of small self-financed urban development projects and with reduced public spending on infrastructure (I. Chorianopoulos 2003; Berg and Braun 2017; Panori et al. 2019). In recent decades, the high population density reached in central urban areas has led to the need to reorganize the structure of the metropolitan city, relating it in all its parts. The growing demand for commercial, industrial, residential and recreational activities has led to an inevitable expansion of the urban area beyond its traditional borders. The areas that have been most affected by this expansion are those with greater accessibility, with a strong concentration of economic enterprises, located mainly in the Messoghia and Thriassian plains (Chorianopoulos et al. 2010).

The expansion process of the Athenian city can be articulated in various phases of demographic and urban development. The first of these periods (1850–1900) was characterized by the constant growth of urban and rural areas, together with a balanced distribution of the population on the territory. In the second period (1900–1940), the development affected the two main urban centers, that of Athens and Piraeus, where the capital attracted industrialization in its surroundings, especially in Piraeus, where the port is still located today commercially. As a result, population growth rates were very high in these two areas. The demographic development of Athens was also due to internal migration during the Second World War, to the subsequent civil war and refugees, immigrants from Turkey (Leontidou 1990). The latter settled in informal, low-density settlements, which initiated the first widespread city trials. The third phase (1950–1980) was characterized by a transformation of the traditional polycentrism of the region, with a decrease in the difference in population density between urban centers and suburbs. The urbanization process of rural areas began around Athens and Piraeus, particularly in the Messoghia plain. Urban–rural depolarization, favored by the development of infrastructures and especially by permissive urban policies, marks the beginning of the fourth phase (1980–1990), characterized by sprawl, in which the exodus to the metropolitan area was significant. The outskirts of Athens (including the municipalities of Acharnes, Ano Liossia, Filis), rural areas (such as Messoghia) and coastal areas experienced rapid urban growth in those years (Salvati and Serra 2016).

Today Athens has a metropolitan area (Fig. 3.9) with small villages and large urban agglomerations. Because of the morphological characteristics of the Greek territory, the urban system is characterized by a strong city-countryside polarization.

Fig. 3.9 Athens cityscape (Photo of the author [June 2016])

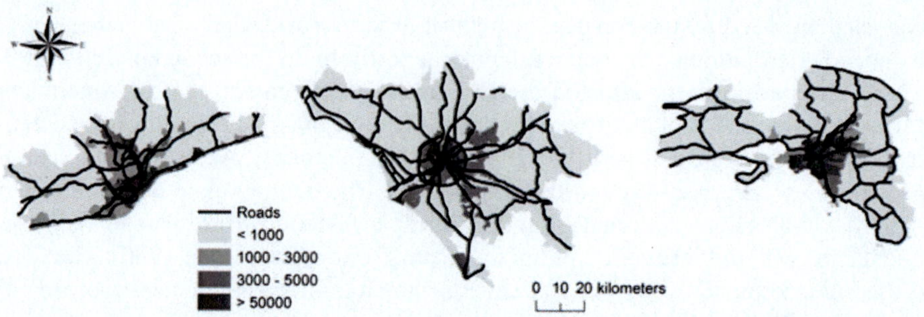

Fig. 3.10 The main road networks and population density (inhabitants per square kilometer) in the cities of Barcelona, Rome, Athens (image modified by Tombolini et al. 2015)

The morphology of a city is usually shaped by the interaction between the biophysical, socio-demographic, economic, cultural and political context. In a comparative analysis, it is, therefore, useful to take into account not only the settlement pattern but also the topography or infrastructure networks. For example, the altitude gradient plays an important role in urban morphology and the expansion of urbanization due to the temperatures, aspect and landscape visibility (Senciales-González et al. 2020). A harsh topography and raised landscapes reduce the availability of building land, promoting more compact morphologies and an urban growth that develops mainly along the coast or in the main plain.

The metropolitan area of Athens has an area that is 80% hilly or mountainous (>200 m high). Rome has expanded into the floodplain of the Tiber River, where the limit to building land is represented by the Apennine mountain range east of the central city. Barcelona is in an intermediate condition, with building land distributed along with the coast and in some inland flat areas in correspondence with river valleys of regional importance.

Spatial analysis of the main road network allows identifying the nuclei of the three cities. In Fig. 3.10, the overlap of the road network is shown on the maps representing the population density. The analysis highlights three different urban forms. In the case of Barcelona, the road network expands connecting to other urban centers in the metropolitan area; for Rome, the ring road concentrates the highest population density, which gradually decreases externally; finally, in Athens, the population density is low in the metropolitan area, growing near Athens and Piraeus, where the roads intersect with each other.

3.4.2 Socio-Economic Aspects

The reference database for the socio-economic characterization of the three cities derives from the portals of Eurostat and the national statistical ones (INE, ISTAT and EL.STAT). Table 3.1 shows some general information, in terms of area and population.

Table 3.1 General characteristics of the metropolitan contexts of Barcelona, Rome and Athens. Processing on Eurostat data

Metropolitan area	Surface (km^2)	Population (2016)	Number of districts (2016)
Barcelona	2.645	4.931.694	75
Rome	6.170	4.414.288	147
Athens	3.040	3.828.434	94

The highest population density is found in the urban areas, which are affected by a high infrastructure of the territory. This density is decreasing from the center to the periphery. The areas surrounding the urban centers of Barcelona and Athens, in particular, have a high density, due to the presence of the main production and commercial settlements at the metropolitan level.

From the maps in Fig. 3.11, it is possible to see how, in general, at the municipal level there has been a considerable increase in the population, compared to a loss

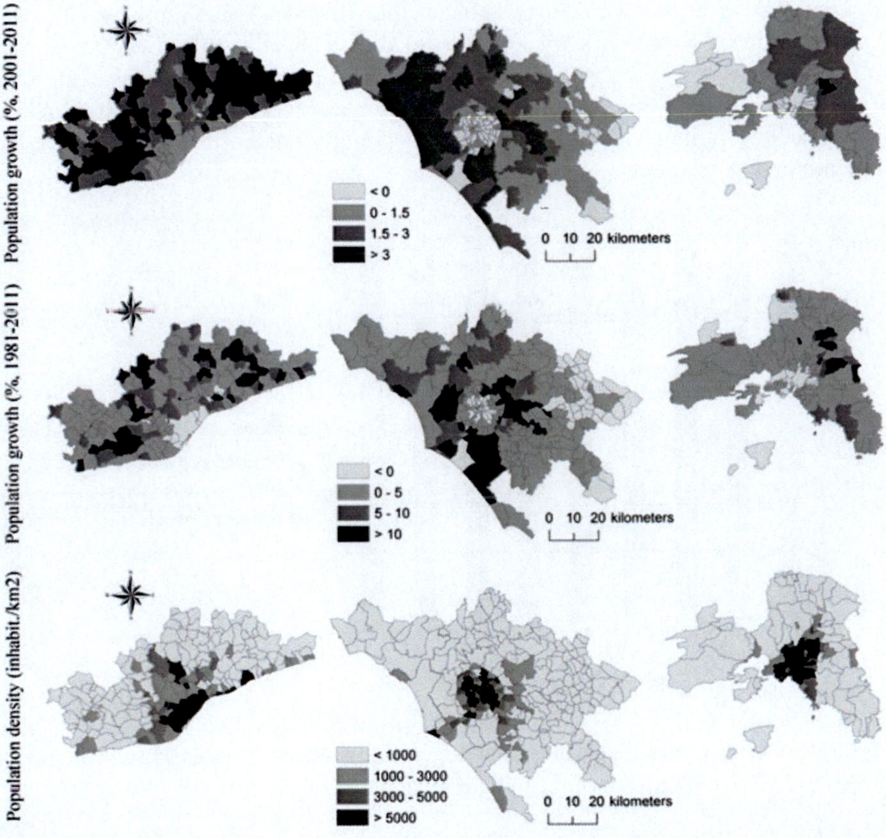

Fig. 3.11 Elaborations on INE, ISTAT and EL.STAT data (image modified by Tombolini et al. 2015)

found in the central areas. Urban sprawl has led to a huge deconcentrating of the flows of people and activities from urban nuclei to peri-urban fringes. One way to evaluate these changes is to consider the temporal variation of population growth. This is a phenomenon that has been observed considering two different time intervals, one of ten years (from 2001 to 2011) and another of thirty years (from 1981 to 2011).

The city of Barcelona and its neighboring municipalities have experienced population growth over the past ten years thanks to the arrival of new inhabitants. If in the long term population growth is very dispersed in suburban areas, the number of residents has increased in the past ten years. The urban growth of Rome has dispersed along the radial axes with respect to the historic center, especially along the streets and creating a fragmented landscape at progressively greater distances from the central districts. This growth can also be seen with the demographic changes in the metropolitan area (Fig. 3.12).

In the past 30 years, the municipalities that have experienced the largest population increase are in the metropolitan area while the Roman city has seen a population loss. The increase in residents mainly affected the municipalities near the coast. In the past ten years, however, the demographic change has been much more pronounced. The urban area of Rome not only continues to show a negative value but also extends to the nearby toponymical subdivisions. Long-term population growth seems very fragmented, while in the recent ten-year period it appears to be continuous and, in particular, the further you move away from the city center, the more there is an increase in new residents.

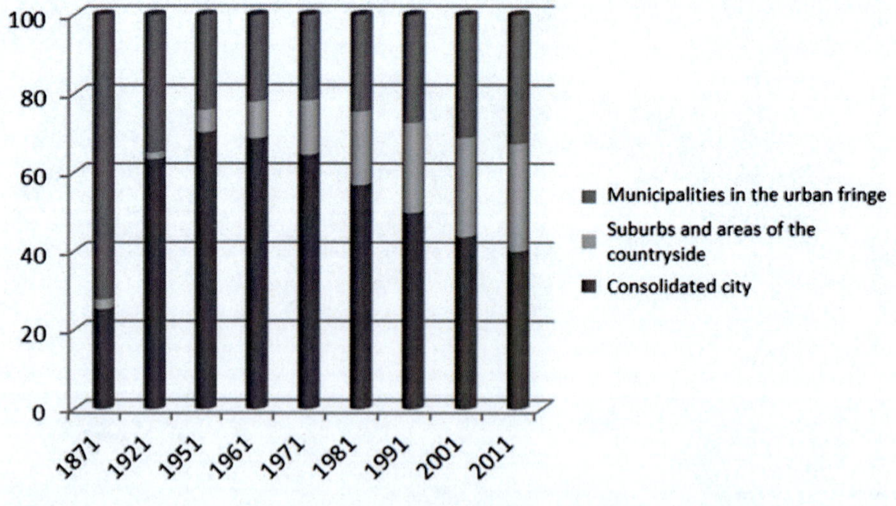

Fig. 3.12 Percentage composition of the population in the province of Rome by zone. The consolidated city includes the toponymical subdivisions of the districts and neighborhoods inherent in the municipality of Rome; the suburbs and areas of the countryside represent the toponymical subdivisions of the peripheral crown, but always belonging to the municipality of Rome; the other municipalities in the province are jointly considered in the last category (*Source* Salvati and Sabbi (2011), with the processing of ISTAT data, General Population Censuses from 1871 to 2011)

As for Athens, in the long term, the growth of the population of its metropolitan area appears with average values, except for the areas near Piraeus, where there has been a loss, and some municipalities in the area to the east (such as Artemida, Pikermiou and Pallini), along the coast, where there are high rates of new residents. Although there has been a moderate demographic change in the long run, in the short run, the entire eastern part has been more populated, with medium-high values of new residents, while going westward, these values become low or even negative. As for the more compact area, the area around Piraeus and the city of Athens had a loss of residents. In the past ten years, the municipalities with the highest growth of new resident population are Pallini, Anthousis (Koinotita) and Gerakas (Salvati 2016b).

3.4.3 *Transforming Urban Europe: The Mediterranean Lesson*

The socio-economic structure of Barcelona, Rome and Athens appears very different, although they are all located in the Mediterranean basin. Each one has a unique and different territorial configuration and the phenomenon of urban sprawl has adapted differently, following the economic and social connotations of the countries under investigation.

Despite a different past, the three metropolitan areas have witnessed the formation of dispersed settlements, fueled by forms of illegality and illegality, which have not taken urban planning into account. The presence of these settlements has led to the creation of socially polarized spaces, homogenization and social segregation (Pacione 2003; Le Goix 2005). The lifestyle of the inhabitants of these contexts is also a consequence of globalization, the idea of living in one's residence having spread in imitation of the American suburbs (Longhi and Musolesi 2007). This figure was found especially in the metropolitan area of Barcelona (Cuadrado-Ciuraneta et al. 2017).

The comparative analysis of the three cities investigated allowed a better definition of the spatial configuration of the respective metropolitan areas, each characterized by a particular economic and social structure (Duvernoy et al. 2018). Although they are all located in the Mediterranean area, they differ in various aspects (Di Feliciantonio et al. 2018; Perrin et al. 2018). Metropolitan areas have undergone the same phenomenon of sprawl, which, however, has mixed well with the social and economic connotations of each context (Biasi et al. 2015a). If in some contexts, economic development together with the phenomena of urban repolarization and densification tends to balance the socio-economic structure on a regional scale. In other contexts, economic disparities seem to be consolidated, also due to the economic crisis and the creation of settlements even more socially polarized (Vidal et al. 2011). All this has led to a different spatial configuration of cities. In fact, the Metropolitan area of Barcelona shows a greater polycentrism, with suburban nuclei having a decisive role in the territory, overall balanced (Garcia-López and Muñiz

2010). In this context, the presence of strong economic poles dispersed throughout the metropolitan area has given rise to a polycentric system that manages to communicate well in every part of the territory, avoiding the proliferation of socially segregated realities. In the case of Barcelona, the birth of a new lifestyle can be confirmed, which is in line with the model of the American suburb. The metropolitan area of Rome appears to be the most chaotic, "scattered" case, because of its deregulated past, highlighting how low-density settlements were born spontaneously (Salvati and Zitti 2011). In this case, the complex system, without a specific spatial organization, and the lack of widespread economic activities in the territory cause a form of the social and institutional disorder, which shows a strong dependence on the capital city. The Greek context, however, with its metropolitan area, appears more monocentric and Athens maintains its consolidated role as a capital city, despite the presence of specific areas intended exclusively for low-density settlements (Kourliouros 1997). In the latter, there is strong social segregation (Leontidou 1993; Maloutas 2007; Couch et al. 2008a), as in the eastern part of Attica rich neighborhoods have spread, with high levels of per capita income, high presence of professional positions, to entrepreneurial example, the high employment rate (Rontos et al. 2016). All this leads to a socio-economic configuration on a regional scale which highlights processes of fragmentation and social and spatial isolation.

References

Allegretti C, Cellamare G (2009) The ambiguous renaissance of Rome. In Whose urban renaissance? an international comparison of urban regeneration strategies, Routledge, London, UK

Amin A, Thrift N (2005) Città. Ripensare la dimensione urbana, Il Mulino, Bologna, Italy

Arapoglou VP, Sayas J (2009) New facets of urban segregation in southern Europe: gender, migration and social class change in Athens. European Urban and Regional Studies, Sage, London, UK. https://doi.org/10.1177/0969776409340187

Arnaiz-Schmitz C, Díaz P, Ruiz-Labourdette D, Herrero-Jáuregui C, Molina M, Montes C, Pineda FD, Schmitz MF (2018) Modelling of socio-ecological connectivity. the rural-urban network in the surroundings of Madrid (Central Spain). Urban Ecosyst 21:1199–1212. https://doi.org/10.1007/s11252-018-0797-z

Bajocco S, De Angelis A, Salvati L (2012) A satellite-based green index as a proxy for vegetation cover quality in a Mediterranean region. Ecol Ind 23:578–587. https://doi.org/10.1016/j.ecolind.2012.05.013

Bajocco S, Salvati L, Ricotta C (2011) Land degradation versus fire: a spiral process? Prog Phys Geogr: Earth Environ 35:3–18. https://doi.org/10.1177/0309133310380768

Bakr N, Weindorf DC, Bahnassy MH, El-Badawi MM (2012) Multi-temporal assessment of land sensitivity to desertification in a fragile agro-ecosystem: environmental indicators. Ecol Ind 15:271–280. https://doi.org/10.1016/j.ecolind.2011.09.034

Berkes F, Folke C, Colding J (2000) Linking social and ecological systems: management practices and social mechanisms for building resilience, Cambridge University Press

Biasi R, Brunori E, Smiraglia D, Salvati L (2015a) Linking traditional tree-crop landscapes and agro-biodiversity in central Italy using a database of typical and traditional products: a multiple risk assessment through a data mining analysis. Biodivers Conserv 24:3009–3031. https://doi.org/10.1007/s10531-015-0994-5

Biasi R, Colantoni A, Ferrara C, Ranalli F, Salvati L (2015b) In-between sprawl and fires: long-term forest expansion and settlement dynamics at the wildland–urban interface in Rome, Italy. Int J Sustain Dev World Ecol 22:467–475. https://doi.org/10.1080/13504509.2015.1064488

Bini P, Crisci M, Gemmiti R, Violante A (2014) Urban sprawl e shrinking cities in Italia. CNR-IRPPS e-Publishing, Sapienza Università di Roma, Italy, Trasformazione urbana e redistribuzione della popolazione nelle aree metropolitane

Blondel J (2006) The "design" of Mediterranean landscapes: a millennial story of humans and ecological systems during the historic period. Hum Ecol 34:713–729. JSTOR.

Carlucci M, Chelli FM, Salvati L (2018) Toward a new cycle: short-term population dynamics, gentrification, and re-urbanization of Milan (Italy). Sustainability 10:1–12

Catalán B, Saurí D, Serra P (2008) Urban sprawl in the Mediterranean? patterns of growth and change in the Barcelona metropolitan region 1993–2000. Landscape Urban Plan 85:174–184

Cecchini M, Zambon I, Pontrandolfi A, Turco R, Colantoni A, Mavrakis A, Salvati L (2019) Urban sprawl and the 'olive' landscape: sustainable land management for 'crisis' cities. GeoJournal 84:237–255. https://doi.org/10.1007/s10708-018-9848-5

Chelli FM, Ciommi M, Emili A, Gigliarano C, Taralli S (2016) Assessing the equitable and sustainable well-being of the Italian provinces. Int J Uncertainty Fuzziness Knowl Based Syst 24:39–62. https://doi.org/10.1142/S0218488516400031

Chelli F, Rosti L (2002) Age and gender differences in Italian workers' mobility. Int J Manpower 23:313–325. https://doi.org/10.1108/01437720210439656

Chorianopoulos I (2003) North-South local authority and governance differences in EU networks. Eur Plan Stud 11:671–695. https://doi.org/10.1080/0965431032000108404

Chorianopoulos I, Pagonis T, Koukoulas S, Drymoniti S (2010) Planning, competitiveness and sprawl in the Mediterranean city: the case of Athens. Cities 27:249–259

Ciommi M, Chelli FM, Carlucci M, Salvati L (2018) Urban growth and demographic dynamics in Southern Europe: toward a new statistical approach to regional science. Sustainability 10:2765. https://doi.org/10.3390/su10082765

Ciommi M, Chelli FM, Salvati L (2019) Integrating parametric and non-parametric multivariate analysis of urban growth and commuting patterns in a European metropolitan area. Qual Quant 53:957–979. https://doi.org/10.1007/s11135-018-0798-2

Ciommi M, Gentili A, Ermini B, Gigliarano C, Chelli FM, Gallegati M (2017a) Have your cake and eat it too: the well-being of the Italians (1861–2011). Soc Indic Res 134:473–509. https://doi.org/10.1007/s11205-016-1450-y

Ciommi M, Gigliarano C, Emili A, Taralli S, Chelli FM (2017b) A new class of composite indicators for measuring well-being at the local level: an application to the equitable and sustainable well-being (BES) of the Italian provinces. Ecol Ind 76:281–296. https://doi.org/10.1016/j.ecolind.2016.12.050

Clemente M, Zambon I, Konaxis I, Salvati L (2018) Urban growth, economic structures and demographic dynamics: exploring the spatial mismatch between planned and actual land-use in a Mediterranean city. Int Plan Stud 23:376–390. https://doi.org/10.1080/13563475.2018.1500277

Colantoni A, Grigoriadis E, Sateriano A, Venanzoni G, Salvati L (2016) Cities as selective land predators? a lesson on urban growth, deregulated planning and sprawl containment. Sci Total Environ 545–546:329–339. https://doi.org/10.1016/j.scitotenv.2015.11.170

Colantoni A, Mavrakis A, Sorgi T, Salvati L (2015) Towards a 'polycentric' landscape? reconnecting fragments into an integrated network of coastal forests in Rome. Rendiconti Lincei 26:615–624. https://doi.org/10.1007/s12210-015-0394-5

Costa FJ, Noble AG, Pendleton G (1991) Evolving planning systems in Madrid, Rome, and Athens. GeoJournal 24:293–303. JSTOR

Couch C, Petschel-Held G, Leontidou L (2008a) Urban sprawl in Europe: landscape, land-use change and policy | Wiley. Wiley-Blackwel, London, UK

Couch C, Leontidou L, Arnstberg K-O (2008b) Introduction: definitions, theories and methods of comparative analysis. In Urban sprawl in Europe, 1–38. Wiley & Sons, Ltd. https://doi.org/10.1002/9780470692066.ch1

Cuadrado-Ciuraneta S, Durà-Guimerà A, Salvati L (2017) Not only tourism: unravelling suburbanization, second-home expansion and "rural" sprawl in Catalonia, Spain. Urban Geogr 38:66–89. https://doi.org/10.1080/02723638.2015.1113806

De Rosa S, Salvati L (2016) Beyond a 'side street story'? Naples from spontaneous centrality to entropic polycentricism, towards a 'crisis city.' Cities 51. Current Research on Cities: 74–83. https://doi.org/10.1016/j.cities.2015.11.025

Delladetsima PM (2006) The emerging property development pattern in Greece and its impact on spatial development. Eur Urban Reg Stud 13:245–278. https://doi.org/10.1177/0969776406065428

Di Feliciantonio C, Salvati L (2015) 'Southern' alternatives of urban diffusion: investigating settlement characteristics and socio-economic patterns in three Mediterranean regions. Tijdschr Econ Soc Geogr 106:453–470. https://doi.org/10.1111/tesg.12102

Di Feliciantonio C, Salvati L, Sarantakou E, Rontos K (2018) Class diversification, economic growth and urban sprawl: evidences from a pre-crisis European city. Qual Quan 52:1501–1522. https://doi.org/10.1007/s11135-017-0532-5

Díaz-Pacheco J, García-Palomares JC (2014) Urban sprawl in the Mediterranean urban regions in Europe and the crisis effect on the urban land development: Madrid as study case. Research Article. Urban Stud Res, November 26, https://doi.org/10.1155/2014/807381

Dura-Guimera A (2003) Population deconcentration and social restructuring in Barcelona, a European Mediterranean city. Cities 20. Part Special Issue: Cities of Spain: 387–394. https://doi.org/10.1016/j.cities.2003.08.004

Duvernoy I, Zambon I, Sateriano A, Salvati L (2018) Pictures from the other side of the fringe: urban growth and peri-urban agriculture in a post-industrial city (Toulouse, France). J Rural Stud 57:25–35

Ferrara A, Salvati L, Sabbi A, Colantoni A (2014) Soil resources, land cover changes and rural areas: towards a spatial mismatch? Sci Total Environ 478:116–122. https://doi.org/10.1016/j.scitotenv.2014.01.040

Forino G, Ciccarelli S, Bonamici S, Perini L, Salvati L (2015) Developmental policies, long-term land-use changes and the way towards soil degradation: evidence from Southern Italy. Scott Geogr J 131:123–140. https://doi.org/10.1080/14702541.2015.1047895

Garcia-López M-À, Muñiz I (2010) Employment decentralisation: polycentricity or scatteration? the case of Barcelona: urban studies. Sage, London, UK. https://doi.org/10.1177/0042098009360229

Gardi C, Panagos P, Van Liedekerke M, Bosco C, De Brogniez D (2015) Land take and food security: assessment of land take on the agricultural production in Europe. J Environ Plan Manag 58:898–912. https://doi.org/10.1080/09640568.2014.899490

Giannakourou G (2005) Transforming spatial planning policy in Mediterranean countries: Europeanization and domestic change. Eur Plan Stud 13:319–331. https://doi.org/10.1080/0365431042000321857

Gibelli, MC, Salzano, E (2006) No sprawl, Alinea, Firenze

Gigliarano C, Chelli FM (2016) Measuring inter-temporal intragenerational mobility: an application to the Italian labour market. Qual Quan 50:89–102. https://doi.org/10.1007/s11135-014-0138-0

Gisladottir G, Stocking M (2005) Land degradation control and its global environmental benefits. Land Degrad Dev 16:99–112. https://doi.org/10.1002/ldr.687

Goix RL (2005) Gated communities: sprawl and social segregation in Southern California. Hous Stud 20:323–343. https://doi.org/10.1080/026730303042000331808

Graovac Matassi V (2014) Contemporary urban changes in Croatia—the case study of Zadar. Critical spaces—contemporary perspectives in urban, spatial and landscape studies. Austria, Vienna, pp 167–178

Haase A, Bernt M, Großmann K, Mykhnenko V, Rink D (2013) Varieties of shrinkage in European cities. European Urban and Regional Studies, Online version, pp 1–17

Hernández A, Miranda M, Arellano EC, Saura S, Ovalle C (2015) Landscape dynamics and their effect on the functional connectivity of a Mediterranean landscape in Chile. Ecol Indic 48:198–206. https://doi.org/10.1016/j.ecolind.2014.08.010

Hubacek K, van den Bergh, JCJM, (2006) Changing concepts of 'land' in economic theory: from single to multi-disciplinary approaches. Ecol Econ 56:5–27. https://doi.org/10.1016/j.ecolecon.2005.03.033

Indovina, F (2009) Dalla città diffusa all'arcipelago metropolitano, Franco Angeli, Milano

Insolera, I (2001) Roma moderna. Un secolo di storia urbanistica 1870–1970, Einaudi, Torino

Kangalawe RYM, Lyimo JG (2010) Population dynamics, rural livelihoods and environmental degradation: some experiences from Tanzania. Env Dev Sustain 12:985–997. https://doi.org/10.1007/s10668-010-9235-y

Kelly C, Ferrara A, Wilson GA, Ripullone F, Nolè A,. Harmer N, Salvati L (2015) Community resilience and land degradation in forest and shrubland socio-ecological systems: evidence from Gorgoglione, Basilicata, Italy. Land Use Policy 46:11–20. https://doi.org/10.1016/j.landusepol.2015.01.026

Kosmas C, Karamesouti M, Kounalaki K, Detsis V, Vassiliou P, Salvati L (2016) Land degradation and long-term changes in agro-pastoral systems: an empirical analysis of ecological resilience in Asteroussia—Crete (Greece). CATENA 147:196–204. https://doi.org/10.1016/j.catena.2016.07.018

Kourliouros E (1997) Planning industrial location in Greater Athens: the interaction between deindustrialization and anti-industrialism during the 1980s. Eur Plan Stud 5:435–460. https://doi.org/10.1080/09654319708720411

Kranjčević J, Hajdinjak S (2019) Tourism urbanization in Croatia. the cases of Poreč in Istria and Makarska in Dalmatia. Südosteuropa 67:393–420. https://doi.org/10.1515/soeu-2019-0028

Krumholz N (1992) Roman impressions: contemporary city planning and housing in Rome. Landscape Urban Plan 22:107–114. https://doi.org/10.1016/0169-2046(92)90015-R

Kurttila M (2001) The spatial structure of forests in the optimization calculations of forest planning—a landscape ecological perspective. Forest Ecol Manage 142:129–142. https://doi.org/10.1016/S0378-1127(00)00343-1

Leontidou L (1990) The Mediterranean City in transition: social change and urban development. Cambridge Human Geography, Cambridge University Press, Cambridge. https://doi.org/10.1017/CBO9780511522208

Leontidou L (1993) Postmodernism and the city: Mediterranean versions. Urban Stud 30:949–965. https://doi.org/10.1080/00420989320080881

Lesthaeghe R, Lopez-Gay A (2013) Spatial continuities and discontinuities in two successive demographic transitions: Spain and Belgium, 1880–2010. Demogr Res 28:77–136. https://doi.org/10.4054/DemRes.2013.28.4

Longhi C, Musolesi A (2007) European cities in the process of economic integration: towards structural convergence. The Ann Reg Sci 41:333–351. https://doi.org/10.1007/s00168-006-0104-4

Maksin-Mićić M (2008) Problems and possibilities for steering urban sprawl in Serbia. In Urban growth without sprawl—a way towards sustainable urbanization, 1–12. Dalian, China.

Maloutas T (2007) Segregation, social polarization and immigration in Athens during the 1990s: theoretical expectations and contextual difference. Int J Urban Reg Res 31:733–758. https://doi.org/10.1111/j.1468-2427.2007.00760.x

Martinotti G (1993) Metropoli. La nuova morfologia sociale della città, Il Mulino, Bologna

Mazzoleni C (2009) La costruzione dello spazio urbano: l'esperienza di Barcellona, Franco Angeli, Milano, Italy

McNeill D (1999) Globalization and the European city. Cities 16:143–148. https://doi.org/10.1016/S0264-2751(99)00012-8

Minca C (2003) Orizzonte mediterraneo, CEDAM, Padova, Italy

Montanarella L (2007) Trends in land degradation in Europe. In: Sivakumar MVK, Ndiang'ui, N (eds) Climate and land degradation. Environmental science and engineering, Springer, Berlin, Heidelberg, pp 83–104. https://doi.org/10.1007/978-3-540-72438-4_5

Morelli VG, Rontos K, Salvati L (2014) Between suburbanisation and re-urbanisation: revisiting the urban life cycle in a Mediterranean compact city. Urban Res Pract 7:74–88. https://doi.org/10.1080/17535069.2014.885744

Morelli VG, Salvati L (2010) Ad hoc urban sprawl in the Mediterranean city: dispersing a compact tradition? Edizioni Nuova Cultura

Morollón FR, González Marroquin VM, Pérez Rivero JL (2016) Urban sprawl in Spain: differences among cities and causes. Euro Plan Stud 24:207–226. https://doi.org/10.1080/09654313.2015.1080230

Mosconi EM, Colantoni A, Gambella F, Cudlinová E, Salvati L, Rodrigo-Comino J (2020) Revisiting the environmental Kuznets curve:the spatial interaction between economy and territory. Economies 8:74. https://doi.org/10.3390/economies8030074

Munafò M, Salvati L, Zitti M (2013) Estimating soil sealing rate at national level—Italy as a case study. Ecol Indic 26:137–140. https://doi.org/10.1016/j.ecolind.2012.11.001

Munoz F (2003) Lock living: urban sprawl in Mediterranean cities. Cities 20. Part Special Issue: Cities of Spain: 381–385. https://doi.org/10.1016/j.cities.2003.08.003

Muscarà C (1978) Megalopoli mediterranea. Franco Angeli, Milano

Myers N, Mittermeier RA, Mittermeier CG, da Fonseca GAB, Kent J (2000) Biodiversity hotspots for conservation priorities. Nature 403:853–858. https://doi.org/10.1038/35002501

Pace G (1996) Modi di pensare e vedere la città mediterranea. Istituto Italiano per gli Studi Filosofic

Pacione M (2003) Urban environmental quality and human wellbeing—a social geographical perspective. Landscape Urban Plan 65. Urban Environmental Quality and Human Wellbeing: 19–30. https://doi.org/10.1016/S0169-2046(02)00234-7

Panori A, Psycharis Y, Ballas D (2019) Spatial segregation and migration in the city of Athens: investigating the evolution of urban socio-spatial immigrant structures. Popul Space Place 25:e2209. https://doi.org/10.1002/psp.2209

Paül V, Tonts M (2005) Containing urban sprawl: trends in land use and spatial planning in the Metropolitan region of Barcelona. J Environ Plan Manage 48:7–35. https://doi.org/10.1080/0964056042000308139

Perrin C, Nougarèdes B, Sini L, Branduini P, Salvati L (2018) Governance changes in peri-urban farmland protection following decentralisation: a comparison between Montpellier (France) and Rome (Italy). Land Use Policy 70:535–546

Pirotte A, Madre JL (2011) Determinants of urban sprawl in France: an analysis using a hierarchical bayes approach on panel data. urban studies. Sage, London, UK. https://doi.org/10.1177/0042098010391303

Polinesi G, Recchioni MC, Turco R, Salvati L, Rontos K, Rodrigo-Comino J, Benassi F (2020) Population trends and urbanization: simulating density effects using a local regression approach. ISPRS Int J Geo-Information 9:454. https://doi.org/10.3390/ijgi9070454

Portnov BA, Safriel UN (2004) Combating desertification in the Negev: dryland agriculture vs. dryland urbanization. J Arid Environ 56:659–680. https://doi.org/10.1016/S0140-1963(03)00087-9

Pozzolo LD (2001) Fuori città, senza campagna. Paesaggio e progetto nella città diffusa. Franco Angeli, Milano, Italy

Quaranta G, Salvia R (2005) Riqualificazione e gestione del territorio, lotta alla desertificazione e sviluppo sostenibile. Buone pratiche per i territori rurali, Franco Angeli, Milano, Italy

Recatalá L, Ive JR, Baird IA, Hamilton N, Sánchez J (2000) Land-use planning in the Valencian Mediterranean region: using LUPIS to generate issue relevant plans. J Environ Manage 59:169–184. https://doi.org/10.1006/jema.2000.0350

Richardson HW (ed) (2014) Shrinking cities: a global perspective. Routledge, London, New York

Rodrigo-Comino J, Giménez-Morera A, Panagos P, Pourghasemi HR, Pulido, M, Cerdà, A (2020) The potential of straw mulch as a nature-based solution for soil erosion in olive plantation treated with glyphosate: a biophysical and socioeconomic assessment. Land Degra Devel 31:1877–1889. https://doi.org/10.1002/ldr.3305

Rontos K, Grigoriadis E, Syrmali S, Vavouras M, Salvati L (2016) Lost in protest, found in segregation: divided cities in the light of the 2015 "Oχι" referendum in Greece. City Cult Soc 7:139–148. https://doi.org/10.1016/j.ccs.2016.05.006

Salvati L (2012) The spatial nexus between population growth and land degradation in a dry Mediterranean region: a rapidly changing pattern? Int J Sustain Dev World Ecol 19:81–88. https://doi.org/10.1080/13504509.2011.593007

Salvati L (2014a) Population distribution and urban growth in Southern Italy, 1871–2011:emergent polycentrism or path-dependent monocentricity? Urban Geogr 35:440–453. https://doi.org/10.1080/02723638.2014.881017

Salvati L (2014b) Agro-forest landscape and the 'fringe' city: a multivariate assessment of land-use changes in a sprawling region and implications for planning. Sci Total Environ 490:715–723. https://doi.org/10.1016/j.scitotenv.2014.05.080

Salvati L (2016a) The dark side of the crisis: disparities in per capita income (2000–12) and the urban-rural gradient in Greece. Tijdschr Econ Soc Geogr 107:628–641. https://doi.org/10.1111/tesg.12203

Salvati L (2016b) From sprawl to compactness and back: population dynamics (1848–2011) and the economic structure of a Mediterranean city. GeoJournal 81:319–332. https://doi.org/10.1007/s10708-015-9621-y

Salvati L (2018) Population growth and the economic crisis: understanding latent patterns of change in Greece, 2002–2016. Lett Spat Resour Sci 11:105–126. https://doi.org/10.1007/s12076-018-0204-7

Salvati L, Carlucci M (2011) The economic and environmental performances of rural districts in Italy: are competitiveness and sustainability compatible targets? Ecol Econ 70:2446–2453. https://doi.org/10.1016/j.ecolecon.2011.07.030

Salvati L, Carlucci M (2016) Patterns of sprawl: the socioeconomic and territorial profile of dispersed urban areas in Italy. Reg Stud 50:1346–1359. https://doi.org/10.1080/00343404.2015.1009435

Salvati L, Carlucci M, Grigoriadis E, Chelli FM (2018a) Uneven dispersion or adaptive polycentrism? Urban expansion, population dynamics and employment growth in an 'ordinary' city. Rev Reg Res 38:1–25. https://doi.org/10.1007/s10037-017-0115-x

Salvati L, Ciommi MT, Serra P, Chelli FM (2019) Exploring the spatial structure of housing prices under economic expansion and stagnation: the role of socio-demographic factors in metropolitan Rome, Italy. Land Use Policy 81:143–152. https://doi.org/10.1016/j.landusepol.2018.10.030

Salvati L, De Rosa S (2014) 'Hidden polycentrism' or 'subtle dispersion'? urban growth and long-term sub-centre dynamics in three Mediterranean cities. Land Use Policy 39:233–243. https://doi.org/10.1016/j.landusepol.2014.02.012

Salvati L, Ferrara A, Chelli F (2018b) Long-term growth and metropolitan spatial structures: an analysis of factors influencing urban patch size under different economic cycles. Geogr Tidsskr-Dan J Geogr 118:56–71. https://doi.org/10.1080/00167223.2017.1386582

Salvati L, Gemmiti R, Perini L (2012a) Land degradation in Mediterranean urban areas: an unexplored link with planning? Area 44:317–325. https://doi.org/10.1111/j.1475-4762.2012.01083.x

Salvati L, Lamonica GR (2020) Containing urban expansion: densification vs greenfield development, socio-demographic transformations and the economic crisis in a Southern European City, 2006–2015. Ecol Indic 110:105923. https://doi.org/10.1016/j.ecolind.2019.105923

Salvati L, Morelli VG (2014) Unveiling urban sprawl in the Mediterranean region: towards a latent urban transformation? In J Urban Reg Res 38:1935–1953. https://doi.org/10.1111/1468-2427.12135

Salvati L, Perini L, Sabbi A, Bajocco S (2012b) Climate aridity and land use changes: a regional-scale analysis. Geogr Res 50:193–203. https://doi.org/10.1111/j.1745-5871.2011.00723.x

Salvati L, Petitta M, Ceccarelli T, Perini L, Di Battista F, Elisa M, Scarascia V (2008) Italy's renewable water resources as estimated on the basis of the monthly water balance. Irrig Drainage 57:507–515. https://doi.org/10.1002/ird.380

Salvati L, Sabbi A (2011) Exploring long-term land cover changes in an urban region of southern Europe. Intl J Sustain Dev World Ecol 18:273–282. https://doi.org/10.1080/13504509.2011.560453

Salvati L, Sateriano A, Grigoriadis E (2016a) Crisis and the city: profiling urban growth under economic expansion and stagnation. Lett Spat Resour Sci 9:329–342. https://doi.org/10.1007/s12076-015-0160-4

Salvati L, Serra P (2016) Estimating rapidity of change in complex urban systems: a multidimensional, local-scale approach. Geogr Anal 48:132–156. https://doi.org/10.1111/gean.12093

Salvati L, Zambon I, Chelli FM, Serra P (2018c) Do spatial patterns of urbanization and land consumption reflect different socioeconomic contexts in Europe? Sci Total Environ 25:722–730. https://doi.org/10.1016/j.scitotenv.2017.12.341

Salvati L, Zitti M (2005) Land degradation in the Mediterranean basin: linking bio-physical and economic factors into an ecological perspective. Biota 5:67–77

Salvati L, Zitti M (2008) Regional convergence of environmental variables: empirical evidences from land degradation. Ecol Econ 68:162–168. https://doi.org/10.1016/j.ecolecon.2008.02.018

Salvati L, Zitti M (2009) Convergence or divergence in desertification risk? scale-based assessment and policy implications in a Mediterranean country. J Environ Plan Manage 52:957–971. https://doi.org/10.1080/09640560903181220

Salvati L, Zitti M (2011) Economic growth vs. land quality: a multidimensional approach in Italy. J Environ Plan Manag 54:733–748. https://doi.org/10.1080/09640568.2010.528612

Salvati L, Zitti M, Ceccarelli T (2007) Integrating economic and environmental indicators in the assessment of desertification risk: a case study. Appl Ecol Environ Res 6:129–138. https://doi.org/10.15666/aeer/0601_129138

Salvati L, Zitti M, Perini L (2016b) Fifty years on: long-term patterns of land sensitivity to desertification in italy. land Degrad Dev 27:97–107. https://doi.org/10.1002/ldr.2226

Santos M, Cabral JA (2004) Development of a stochastic dynamic model for ecological indicators' prediction in changed Mediterranean agroecosystems of north-eastern Portugal. Ecol Indic 3:285–303. https://doi.org/10.1016/j.ecolind.2003.11.007

Saurí D (2003) Lights and shadows of urban water demand management: the case of the Metropolitan Region of Barcelona. Eur Plan Stud 11:229–243. https://doi.org/10.1080/09654310303639

Schneider A, Woodcock CE (2008) Compact, dispersed, fragmented, extensive? a comparison of urban growth in twenty-five global cities using remotely sensed data, pattern Metrics and Census Information: Urban Studies Sage. London, UK. https://doi.org/10.1177/0042098007087340

Senciales-González JM, Rodrigo-Comino J, Smith P (2020) Surveying topographical changes and climate variations to detect the urban heat island in the city of Málaga (Spain). Cuadernos de Investigación Geográfica 46:521–543. https://doi.org/10.18172/cig.4228

Sereni E (1996) Storia del paesaggio agrario italiano, Laterza, Bari

Seronde-Babonaux AM (1983) Roma: dalla città alla metropoli, Editori Riuniti.

Seta PD, Della Seta R (1988) I suoli di Roma: uso e abuso del territorio nei cento anni della capitale, Editori Riuniti, Roma, Italy

Shaw BJ, van Vliet J, Verburg PH (2020) The peri-urbanization of Europe: a systematic review of a multifaceted process. Landscape Urban Plan 196:103733. https://doi.org/10.1016/j.landurbplan.2019.103733

Siciliano G (2008) Social multicriteria evaluation of farming practices in the presence of soil degradation. a case study in Southern Tuscany, Italy. Environ Dev Sustain 11:1107. https://doi.org/10.1007/s10668-008-9169-9

Slaev AD, Nedović-Budić Z, Krunić N, Petrić J, Daskalova D (2018) Suburbanization and sprawl in post-socialist Belgrade and Sofia. Eur Plan Stud 26:1389–1412. https://doi.org/10.1080/09654313.2018.1465530

Tombolini I, Salvati L, Sateriano A (2017) Esplorando la complessità urbana. Kappa editore, Roma

Torrens PM (2008) A toolkit for measuring sprawl. Appl Spat Anal Policy 1:5–36. https://doi.org/10.1007/s12061-008-9000-x

Trisorio A (2005) Misurare la sostenibilità. Indicatori per l'agricoltura italiana, 449. Istituto Nazionale di Economia Agraria, Roma, Italy

Uršič M (2012) Urbanisation history of Slovenia in the context of eastern and south- eastern European countries. Eur Q Polit Attitudes and Mentalities 1:38–47

van den Berg L, Braun E (2017) National policy responses to urban challenges in Europe. Ashgate, Burlington

Vidal M, Domene E, Sauri D (2011) Changing geographies of water-related consumption: residential swimming pools in suburban Barcelona. Area 43:67–75. https://doi.org/10.1111/j.1475-4762.2010.00961.x

Zambon I, Benedetti A, Ferrara C, Salvati L (2018) Soil matters? a multivariate analysis of socioeconomic constraints to urban expansion in Mediterranean Europe. Ecolo Econ 146:173–183. https://doi.org/10.1016/j.ecolecon.2017.10.015

Zambon I, Rontos K, Serra P, Colantoni A, Salvati L (2019) Population dynamics in Southern Europe: a local-scale analysis, 1961–2011. Sustainability 11:109. https://doi.org/10.3390/su11010109

Zambon I, Serra P, Sauri D, Carlucci M, Salvati L (2017) Beyond the 'Mediterranean city': socioeconomic disparities and urban sprawl in three Southern European cities. Geogr Annaler: Series b, Human Geogr 99:319–337. https://doi.org/10.1080/04353684.2017.1294857

Chapter 4
What Type of Soil Was Consumed in the Metropolis of the Mediterranean Area? Land Quality and the Forms of Urbanization

Abstract Mediterranean European cities have undergone a transition from compact growth to a more discontinuous and dispersed spatial pattern during the last decades. It is characterized by the irregular expansion of low-density settlements. In many urban areas, the expansion of compact settlements first consumed low-quality soils and moderately degraded landscapes (pastures, abandoned fields and low-intensity agricultural areas) bordering large cities. Also, a progressive increase of the consumption of fertile and in good environmental condition agricultural land has been observed, more and more distant from the urban nuclei, as a result of the sprawl not only causing the fragmentation of natural ecosystems and semi-natural, but also deteriorating the productive capacity and potential of the agrosystems, and the esthetical value of the rural landscape. Representing these dynamics a serious threat to the cohesion and stability of local communities as well as to the quality and diversity of the landscapes. In this chapter, we explore the link that exists between the spread of urbanized soil and the context in which it occurs, investigating how the various forms of urban expansion affect land quality at the metropolitan scale. This exploratory analysis will be treated in the following sub-paragraphs, illustrating the methodology, the study area and the results that emerged.

Keywords Land consumption · Land quality · Urban expansion · Survey tools

4.1 The Link Between Forms of Urban Expansion and *Land Quality*

As anticipated in the previous chapters, European cities in the Mediterranean area in recent decades have undergone a transition from compact growth to a more discontinuous and dispersed spatial pattern characterized by the irregular expansion of low-density settlements (Salvati 2014b). In many urban areas, the expansion of compact settlements first consumed low-quality soils and moderately degraded landscapes (pastures, abandoned fields and low intensity agricultural areas) bordering large cities (Barbero-Sierra et al. 2013). On the contrary, in the last decades, a progressive increase of the consumption of fertile and in good environmental condition agricultural land has been observed, more and more distant from the urban nuclei, as a result

I. Tombolini et al., *Land Quality and Sustainable Urban Forms*, Springer Geography,
https://doi.org/10.1007/978-3-030-94732-3_4

of the sprawl (Salvati 2014c), not only causing the fragmentation of natural ecosystems and semi-natural, but also deteriorating the productive capacity and potential of the agrosystems (Santos and Cabral 2004),[1] and the esthetical value of the rural landscape (Biasi et al. 2015a).

Urbanization has played and continues to play a crucial role in the conversion of land (Serra et al. 2014), especially at the expense of the best soils. The decoupling of urban growth from demographic growth in developed countries (which results in the expansion of settlement areas with a stable or declining population) has led to an increase in land consumption progressively distant from urban nuclei (Coisnon et al. 2014).

Representing these dynamics is a serious threat to the cohesion and stability of local communities as well as to the quality and diversity of the landscapes (Biasi et al. 2015b). It is probably useful to explore the link that exists between the spread of urbanized soil and the context in which it occurs, investigating how the various forms of urban expansion affect land quality at the metropolitan scale (Salvati et al. 2018d). This exploratory analysis will be treated in the following sub-paragraphs, illustrating the methodology, the study area and the results that emerged.

4.2 Survey Tools

The link between the spread of urbanization and land quality was explored by intersecting the data on land use with those relating to the quality of the soil of the main metropolises of the study area. Land use has been classified based on the Urban Atlas nomenclature, a product of the "Copernicus Land Monitoring" service, interoperable and freely accessible by any citizen or organization in the world, which derives from a joint initiative by the Directorate General for Policy regional and urban planning of the European Commission and the Directorate General for Enterprise and Industry within the framework of the Copernicus program of the European Union (Salvati and Serra 2016).

The European Earth Observation Program COPERNICUS, previously known as GMES (Global Monitoring for Environment and Security), is a complex set of systems that collects information from multiple sources, i.e. Earth observation satellites and land, sea and airborne sensors. It integrates and processes all this information, providing users, institutional and related to the industry sector, with reliable and updated information through a series of services that concern the environment, the

[1] By agrosystem or agroecosystem we mean a highly anthropized terrestrial ecosystem, whose dynamics, although basically taking place according to the laws of ecology, are artificially controlled and aimed at the production of biomass and energy to be used for economic purposes. The interference of the anthropic factor with the dynamics within the agrosystem is manifested by the control of the composition of the biocoenosis, the physical environmental factors, the flow of energy and matter, and takes the form of agricultural technique in the broad sense. The role of anthropic intervention is to maintain a situation of strong imbalance, fundamental for obtaining significant economic production (Ferrari et al. 2003).

territory and safety. Copernicus also has among its objectives to guarantee Europe substantial independence in the collection and management of data on the state of health of the planet, supporting the needs of European public policies through the provision of precise and reliable services. This information relates to six thematic areas: land, sea, atmosphere, climate change, emergency management and security (Zambon et al. 2018).

The "Copernicus Land Monitoring Service" deals with the field of environmental terrestrial applications, which has four main components: global, pan-European, local, data and reference images. The local component focuses and provides detailed information on sensitive areas and predisposed to changes from an environmental point of view (hotspots). Among these products, URBAN ATLAS derives from this thematic mapping process, providing a comparable interpretation of European land use and coverage data, referring to the last years of the 1990s or early 2010s, at a scale of 1:10,000. In summary, the product input data are as follows:

- satellite images with a spatial resolution of 2.5 m;
- topographic and cartographic maps at different scales (1: 50.000 or more than 1: 50.000);
- the degree of waterproofing for the classes of residential urban fabric based on the specifications of the high-resolution information layer relating to soil sealing, always developed within the framework of the Copernicus program;
- additional ancillary data (local maps, Google Earth, Bing, etc.).

Urban Atlas was created for urban areas with a resident population greater than 100,000 inhabitants, the so-called FUNCTIONAL URBAN AREAS (FUAS). The information layers available were produced for 2006, for 2012 and the 2006–2012 changes. The 2006 information layer relates to 319 FUAs, while the 2012 one initially included the 319 UA 2006 and 374 new FUAs; following a review, some were added and others removed, for 785 FUAs. The soil classification is based on the Corine Land Cover nomenclature and, in particular, it includes 17 "urban" classes with a minimum cartographic unit of 0.25 hectares and 3 "non-urban" with a minimum cartographic unit of 1 hectare. For this study, they have been grouped into nine main land-use categories (Table 4.1): (i) dense urban fabric; (ii) mixed urban fabric; (iii) discontinuous urban fabric; (iv) settlements service; (v) transportation infrastructure; (vi) open spaces with urban uses (e.g. airports, construction sites, land currently unused or awaiting urbanization); (vii) urban green; (viii) agricultural areas; and, (ix) wooded. The threshold levels for distinguishing the dense, mixed and discontinuous urban fabric were set, respectively, in more than 50%, between 30 and 50% and less than 30% of the incidence of waterproofed soil.

By superimposing on this information layer the qualitative–quantitative data on the soil collected with the ESA method (from English Environmentally Sensitive Area), widely used in the Mediterranean area (e.g. Basso et al. 2000; Bajocco et al. 2012; Colantoni et al. 2018), we want to provide a metropolitan scale comparative assessment of land quality and soil vulnerability to degradation, in particular through the soil degradation index (SDI). Given that the purpose of this research is to investigate whether recent urbanization occurred

Table 4.1 The Urban Atlas (UA) land-use classification system adopted in this study

UA code	Description	Conversion code	Description
1110	Continuous urban fabric (soil sealing level or L.S. > 80%)	1	Dense urban fabric (L.S. > 50%)
1121	Dense discontinuous urban fabric (L.S. 50–80%)		
1122	Discontinuous medium density urban fabric (L.S. 30–50%)	2	Mixed urban fabric (L.S. 30–50%)
1123	Discontinuous low-density urban fabric (L.S. 10–30%)	3	Discontinuous urban fabric (L.S. < 30%)
1124	Discontinuous very-low-density urban fabric (L.S. < 10%)		
1130	Isolated structures		
1210	Industrial, commercial, public, military and private units	4	Service settlements
1221	Fast transit roads and associated territories	5	Network for transportation
1222	Other roads and associated territories		
1223	Railways and associated territories		
1230	Ports	6	Open spaces with urban uses
1240	Airports		
1310	Mining sites		
1330	Sites under construction		
1340	Territories without current use		
1410	Urban green areas	7	Urban green spaces, waters and wetlands
1420	Sports and leisure facilities		
5000	Waters and wetlands		
2000	Agricultural areas	8	Agriculture
3000	Forests	9	Forests

at the expense of high-quality soils (Recanatesi et al. 2016). SDI was not calculated for urban areas already consolidated in the late 1990s, but only for areas of recent development (2000s) and those with non-urban land (agricultural, forestry).

The SDI index, developed as part of the DISMED project, jointly assesses the level of soil quality and its sensitivity to degradation (Salvati 2014a). As shown in Fig. 4.1, the SDI varies from 1 (the value that corresponds to the maximum level of land quality and the minimum level of land degradation) to 2 (which, conversely, corresponds to the lowest level of land quality and the highest of land degradation). The SDI raster data used in the analysis are those produced in 2003 and updated in

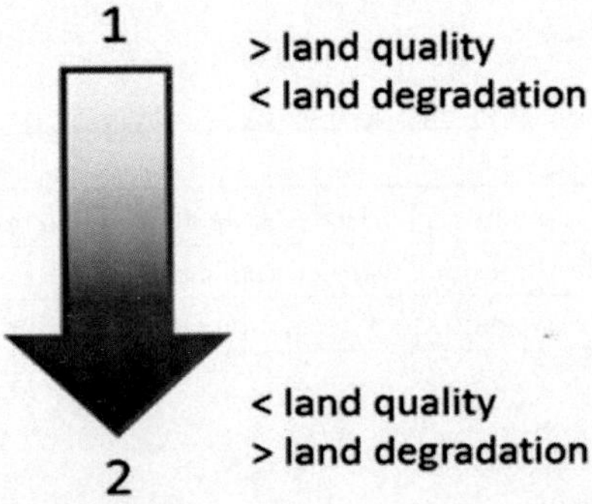

Fig. 4.1 Limit values within which the SDI index can vary. The best soils—more fertile and less degraded—have an SDI close to 1 (maximum level of land quality and a minimum of land degradation); those with an SDI closer to 2 (minimum level of land quality and maximum land degradation) are the least fertile and most degraded soils

2008, available and freely accessible via the website of the European Environment Agency. The methodology of how the product was derived, which derives from the collaboration of the EEA with the ETC-TE (European Topic Center on Terrestrial Environment), is completely analogous to that developed within the MEDALUS project. In detail, the information layer (resolution of 1 km^2 and homogeneously covers the European Mediterranean region), derives from the geometric average of the quality indices relating to soil, climate and vegetation, which are also available and freely accessible through the site of the European Environment Agency:

$$SDI = (SQI \times CQI \times VQI)^{1/3}$$

where:

- SQI = soil quality index;
- CQI = climate quality index;
- VQI = vegetation quality index.

4.2.1 Soil Quality Index

Developed by OSS (Sahara and Sahel Observatory) in 2003, it is given by the geometric mean of four parameters:

- Pedogenetic substrate

Classification (Score)	
Coherent rocks: limestone, dolomite, non-friable sandstone, a layer of hard limestone	(1)
Moderately consistent rocks: limestone marble, friable sandstone	(1.5)
Crumbly rocks: limestone clay, clay, sandy formation, alluvial sediment	(2)

- Soil depth

Class	Description	Score
Very deep	Soil thickness greater than 1.2 m with a substrate that cannot penetrate from the roots or with a thickness greater than 1 m on a mobile substrate	1
Moderate to profound	Depth from 0.8 to 1.2 m with a coherent substrate or 0.5 to 1 m on a mobile substrate	1,33
Shallow	Depth 0.5 to 0.8 m with a coherent substrate or 0.3 to 0.5 m on a mobile substrate	1,66
Very thin	Depth less than 0.3 m	2

Since this type of information has not been collected so extensively, the following approach is adopted:

Pedogenetic substrate	Score
Soils with limited root growth (leptosol)	1
Soils with little or no profile differentiation (fluvisol)	2
All other groups	1,5

- Soil texture

Class	Description	Score
Thin texture	Land silty-sandy, sandy-loam	1
Thin texture compared to the average	Clay soil	1,33
Not very thin to medium texture	Clay soil	1,66
Coarse texture	Sandy to very sandy soil	2

- Soil slope

Class	Description	Score
A	Leveled (slope ranging from 0 to 8%)	1
B	Pending (slope ranging from 8 to 15%)	1,33
C	Moderately steep (slope ranging from 15 to 25%)	1,66
D	Steep (more than 25%)	2

In the case where local data are not available for a single parameter, only the rest are used.

4.2.2 Climate Quality Index

The climatic aspects were analyzed through the aridity index, based on the methodology developed by the FMA (Applied Meteorological Foundation) and derived from the formula:

$$\mathrm{AI} = \frac{P}{\mathrm{PET}}$$

where:

- AI = dryness index;
- P = average annual rainfall;
- PET = mean annual potential evapotranspiration.

Based on this index, the territory is classified as shown as follows:

AI	Climate zone	Classification
<0,05	Very arid	2
0,05–0,20	Arid	1,75
0,20–0,50	Semi-arido	1,50
0,50–0,65	Dry, sub-wet	1,25
>0,65	Wet	1

4.2.3 *Vegetation Quality Index*

This index is calculated from a reclassification of the Corine Land Cover 2000 to the third level, considering for each class of Corine cartography these four parameters: erosion protection; drought resistance; land cover; and fire resistance (Bajocco et al. 2011, 2012, 2016).

The same range of values is used for all four parameters: 1 (good state); 1.5 (moderate state); 2 (bad state). The index, therefore, derives from the geometric average of these four parameters.

To assess the selective consumption of soils with high land quality, the average SDI of the reference land-use class was compared to the average SDI at the landscape scale. For example, if the SDI calculated for a given land-use class is lower than the SDI referred to the landscape; it indicates that urbanization processes consume higher quality soils than the average quality of the building land. The superimposition of the SDI raster on the Urban Atlas map has made it possible to derive information not only on the influence of forms of urban expansion on land quality on the metropolitan scale but, as will be seen in the results, also other indicators on the national scale. In particular, the percentage composition of the land-use classes and four environmental indicators were calculated: (i) the overall surface area of metropolitan areas by country; (ii) the average SDI of metropolitan areas by country; and, (iii) and (iv) two landscape diversity indices calculated on the area as a percentage of the land-use classes considered (Shannon diversity index (H') and Pielou (J) fairness index).

4.3 Study Area

The study area is part of Europe sensitive to land degradation, which overlooks the Mediterranean.

It includes the European countries of Annex IV of the Convention against desertification which fall in the northern part of the Mediterranean basin—Portugal, Spain, Italy, Greece—and the coastal strip of France which overlooks the Mediterranean. Southern Europe is characterized by a wavy topography with vegetation, pedology and characteristic climatic zones that reflect specific factors related to the place and that shape multiple socio-economic contexts. The countries covered by this study share similar environmental characteristics, the so-called temperate climate regime typical of the northern Mediterranean region, specific geological conformations and vegetation types well adapted to aridity and persistent drought periods (Montanarella 2007). The countries considered are characterized by significant socio-economic disparities, such as Italy (northern regions vs southern regions) or Spain (more accessible urban regions vs less accessible rural districts). Territorial disparities may reflect differences in land quality, land use and landscape fragmentation processes (Zambon et al. 2017).

Fig. 4.2 Map of southern Europe with the 76 metropolitan areas considered in this study

For the metropolitan scale analysis, 76 FUAs were selected (7 from Portugal, 22 from Spain, 6 from southern France, 32 from Italy and 9 from Greece), with a resident population consisting of more than 100,000 inhabitants (Fig. 4.2).

Their borders derive from the European Urban Areas statistical classification developed within the Urban Audit program, which identifies the "cities" and their "commuting zones" (constituting the FUNCTIONAL URBAN AREAS or FUAS) and allows the collection of data homogeneous for cities with different socio-economic and environmental characteristics (Salvati et al. 2018b).

The new definition of CITY according to Eurostat (2015) is based on the presence of an urban center, a new spatial concept based on high population density cells, and is divided into four phases (Fig. 4.3):

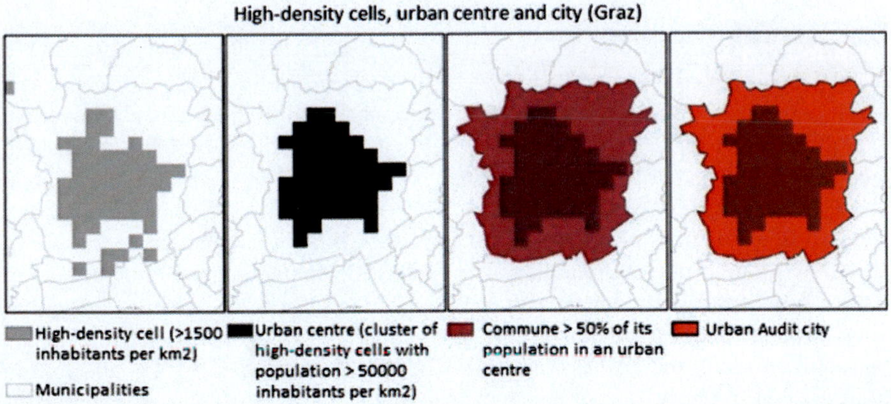

Fig. 4.3 The identification phases of cities, according to the Eurostat definition

- phase 1: all cells with a population density greater than or equal to 1,500 inhabitants per square kilometer are selected;
- phase 2: the high-density contiguous cells are grouped, including any gaps, and only clusters with more than 50,000 inhabitants are maintained as an "urban center";
- phase 3: all municipalities with at least half of the population within the urban center become candidates to be part of the city;
- phase 4: the city is defined ensuring that:
  - there is political correspondence;
  - at least 50% of the population lives in the urban center;
  - at least 75% of the urban center population lives in the city.

Even the COMMUTING ZONES, based on commuting models, have been defined through steps, which are the following:

- if 15% of people with a job, who live in one city, work in another city, they are treated as a city;
- all the municipalities in which at least 15% of their residents with an occupation work in a city are selected;
- municipalities surrounded by a single functional area are included, discarding non-contiguous ones.

The identification of the FUAs represents an attempt to harmonize the definition of "metropolitan area", adopting the concept of the functional urban area, in which a significant proportion of residents move toward the city (Ciommi et al. 2019). A table is presented below (Table 4.2), which summarizes the most updated data on the population and surface of the 76 FUAs considered, and a series of maps (Fig. 4.4), which show the waterproofing of the soil in some of the most representative.

4.4 Results

Overall, in the metropolitan areas investigated, agricultural areas cover on average about 65% of the surface, ranging from 58.1% in French cities to 66.8% in Spanish ones. The wooded areas extend on average for about 20% of the metropolitan area, ranging from 18.5% of Spanish cities to 21.6% of Italian ones. Urban areas occupy about 15% of the metropolitan area, varying from 14% in Spanish cities to 19% in French and Italian ones (Table 4.3). The highest percentage of the compact urban fabric was observed in Italian, French and Greek cities (around 6%), while Portuguese cities have the highest percentage of discontinuous urban fabric (1.8%). At the metropolitan scale, the SDI is high in Spain (1.21) and Greece (1.20) and low in France (1.13) and Italy (1.15). This means that, in general, Spanish and Greek cities have a lower land quality than French and Italian cities due probably to different environmental conditions at the local scale. The comparative analysis of the SDI at

Table 4.2 List of functional urban areas considered in the study, with relative surface area and the most updated population data (2013–2017)

Nation	City	Area	Population
France	Aix En Provence	1206.52	472,736
	Ajaccio	939.24	50,634
	Marseilles	317,87	836,404
	Montpellier	453.21	324,580
	Nice	230.37	392,073
	Toulon	166.11	264,890
Italy	L'Aquila	1531.23	48,883
	Ancona	364.35	223,946
	Bari	810.04	751,181
	Bologna	1905.14	773,511
	Brescia	429.97	479,851
	Cagliari	1471.07	489,215
	Campobasso	1275.56	101,498
	Caserta	572.16	124,500
	Catania	429.29	658,957
	Catanzaro	727.33	156,508
	Cremona	611.81	129,096
	Florence	1140.90	806,237
	Foggia	1023.32	175,576
	Genoa	797.06	720,214
	Milan	875.67	5,097,548
	Modena	577.63	367,301
	Naples	347.94	3,421,906
	Padua	820.30	534,033
	Palermo	1041.10	1,036,973
	Perugia	683.58	282,060
	Pescara	631.63	240,804
	Power	1476.53	130,585
	Reggio di Calabria	396.48	222,189
	Rome	3015.97	4,415,586
	Salerno	844.22	253,559
	Sassari	1059.55	217,045
	Taranto	1147.05	421,542
	Turin	1650.17	1,776,553
	Trento	721.86	235,740

(continued)

Table 4.2 (continued)

Nation	City	Area	Population
	Trieste	124.41	236,073
	Venice	578.99	563,449
	Verona	1049.30	515,838
Spain	Alicante	597.45	462,008
	Badajoz	1438.26	180,542
	Barcelona	1352.13	4,913,865
	Bilbao	895.49	1,038,998
	Cordova	1177.13	306,053
	Gijón	484.66	299,684
	Logroño	1395.19	192,538
	Madrid	7243.23	6,643,994
	Rum raisin	855.16	853,516
	Murcia	1274.20	619,519
	Oviedo	2305.72	312,501
	Palma de Mallorca	1985.80	670,128
	Pamplona	4329.80	376,840
	Santander	544.82	383,014
	Santiago de Compostela	1328.20	199,924
	Zaragoza	2182.67	753,884
	Seville	2759.62	1,542,162
	Toledo	3567.61	149,312
	Valencia	1244.12	1,717,473
	Valladolid	2977.00	424,907
	Vigo	1302.91	543,034
	Vitoria -Gasteiz	2234.63	270,472
Greece	Athens	2418.52	3,828,434
	Kalamata	421.77	69,849
	Candia	586.51	318,092
	Giannina	1285.13	132,979
	Kavala	331.54	70,501
	Larissa	1501.82	195,120
	Patras	442.28	217,555
	Thessaloniki	1253.01	973,997
	Flight	266,60	137,630
Portugal	Aveiro	187.30	139,788

(continued)

Table 4.2 (continued)

Nation	City	Area	Population
	Braga	455.95	247,516
	Coimbra	1218.83	268,783
	Lighthouse	413.57	116,920
	Lisbon	959.75	2,810,668
	Port	408.78	1,279,587
	Setubal	138.77	125,000

the land-use class level and on a national scale allowed to identify mixed spatial patterns of land quality based on the type of urbanization (dense, mixed, dispersed).

Compact urban settlements have expanded to areas with above-average soil quality in the Italian and Portuguese cities; the discontinuous urban fabric has expanded respectively on soils with higher quality than the average SDI in French, Spanish and Portuguese cities.

For the mixed urban fabric, a different pattern was observed as this class occupied soils with a lower quality (compared to the average SDI) in France, Greece, Portugal and slightly higher in Spain (Table 4.3). Soil quality increases with the diversity of the landscape. The analysis of how the soil was consumed according to the type of urban use (Table 4.4) indicates that the discontinuous residential settlements were built on soils with higher quality than the average quality of the soils in the respective metropolitan area in 17 cities out of 76, which make up 22.3% of cases. This percentage falls in the case of areas occupied by service settlements (18.4%), dense urban fabric (17.1%), mixed residential settlements (14.5%) and transport infrastructures (4.0%). Considering all the classes together, selective soil consumption was observed in 40 metropolitan areas out of 76 (52.6%).

In general, discontinuous residential settlements consumed higher quality soils than dense residential settlements (Table 4.5) in 18 out of 76 metropolitan areas (23.7%) which include intermediate and large coastal FUAs in Greece (e.g. Athens, Thessaloniki, Volo), Spain (e.g. Barcelona, Valencia, Seville, Alicante and also Madrid), France (Marseille and Montpellier), Italy (Palermo and Cagliari) and Portugal (Faro, Setubal).

Mixed residential settlements consumed higher quality soils than dense ones in 12 out of 76 cities (15.8%), including Athens and Thessaloniki (Greece), Barcelona, Seville, Alicante, Toledo and Valencia (Spain), Sassari and Genoa (Italy) and Aveiro (Portugal). These cities also have a low to medium level of land quality at the metropolitan scale (Salvati et al. 2018c).

These results indicate that discontinuous urban settlements have consumed higher quality soils in several large and medium-sized metropolitan areas in Greece and Spain. These countries show an overall lower than average soil quality (Barbero-Sierra et al. 2013). Spain and Greece are also the countries in which the sensitivity of lands to degradation is higher (e.g. Wilson and Juntti 2005). In contrast, urban infrastructure and dense residential and service settlements have consumed soils

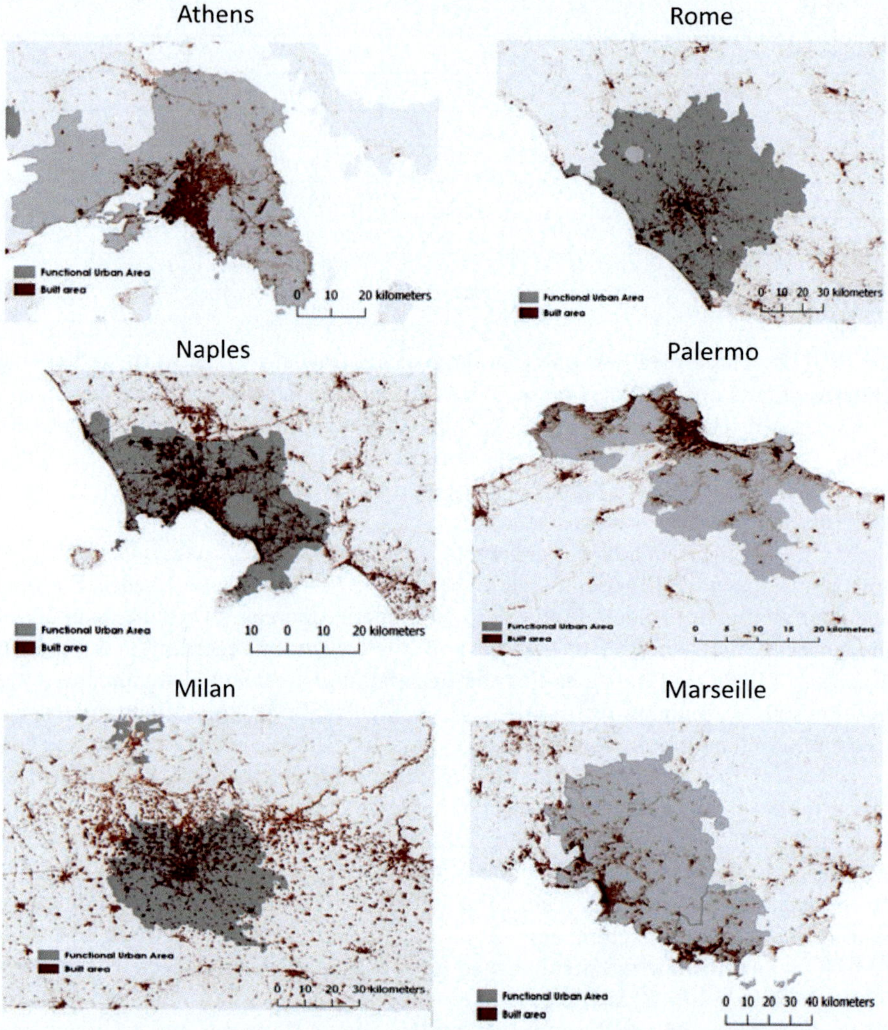

Fig. 4.4 (part 1). High-resolution maps, elaborated on the raster data of the European Environment Agency, which show the waterproofed soil in some of the most representative cities investigated (or FUAs). The product of the European Agency, which derives from the processing of satellite images, consists of a high-resolution information layer (20 m) of the waterproofed soils (2 classes: Built-up and Non-built-up areas). (part 2). High-resolution maps, elaborated on the raster data of the European Environment Agency, which show the waterproofed soil in some of the most representative cities investigated (or FUAs). The product of the European Agency, which derives from the processing of satellite images, consists of a high-resolution information layer (20 m) of the waterproofed soils (2 classes: Built-up and Non-built-up areas)

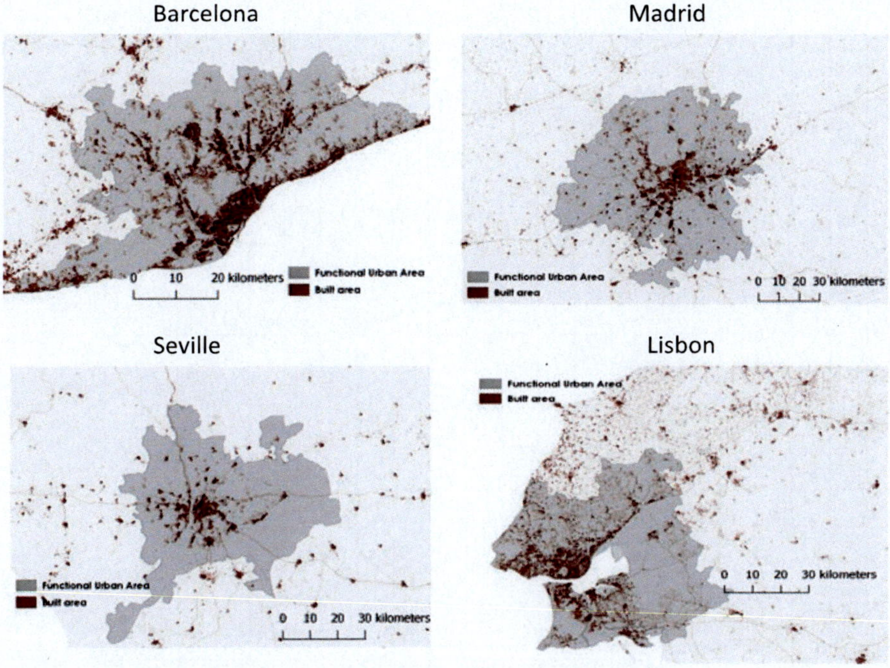

Fig. 4.4 (continued)

with high quality less frequently than the average soil quality on a metropolitan scale (Cuadrado-Ciuraneta et al. 2017; Di Feliciantonio et al. 2018; Lamonica et al. 2020). The dense and mixed settlements, therefore, represent a land-saving model of urban expansion in the rapidly growing cities of southern Europe (Aguilera et al. 2011; A. Ferrara et al. 2014a; Karamesouti et al. 2015).

Discontinuous settlements have selectively consumed high-quality soils in large metropolitan areas (e.g. Barcelona, Madrid, Athens, Thessaloniki or Palermo) or in smaller ones where land quality is low or very low due to the joint action of conditions biophysics, including poor soils and climatic aridity, as in Greece, central-southern Spain and southern Italy. These results confirm the evidence provided by Salvati et al. (2012) and Barbero-Sierra et al. (2013), respectively for Italy and Spain, and indicate the key role of deregulated urban sprawl and building speculation in the increased consumption of land with high quality (Portnov and Safriel 2004). Selective consumption, led by sprawl, of high-quality lands has produced a spatial division between forestry and agricultural uses and built-up areas which progressively occupy the most productive and fertile agricultural land, with negative consequences for the sustainability of the primary sector (Ferrara et al. 2014b), the conservation of natural environments and ecosystem services (Lafortezza et al. 2009).

Table 4.3 Descriptive statistics of the indicators selected by country

Variable	France	Greece	Italy	Portugal	Spain
No. metropolitan areas	6	9	33	7	22
SDI by land-use class					
Dense urban fabric	1,126	1,199	1,149	1,170	1,210
Mixed urban fabric	1,128	1,205	1,151	1,177	1,208
Discontinuous urban fabric	1,123	1,199	1,151	1,163	1,209
Service settlements	1,128	1,198	1,150	1,178	1,210
Transport infrastructure	1,127	1,198	1,152	1,171	1,211
Open areas with urban uses	1,122	1,198	1,147	1,175	1,204
Urban green spaces, waters, wetlands	1,129	1,200	1,152	1,175	1,213
Agriculture	1,128	1,201	1,152	1,167	1,210
Forests	1,127	1,194	1,151	1,173	1,211
Area in percentage by a class of land use					
Dense urban fabric	5.8	5.6	5.8	5.0	3.9
Mixed urban fabric	2.9	2.7	2.7	2.6	2.1
Discontinuous urban fabric	1.5	1.6	1.6	1.8	1.6
Service settlements	3.7	3.8	4.0	3.5	2.8
Transport infrastructure	3.3	3.1	3.0	2.7	2.3
Open areas with urban uses	1.4	1.5	1.6	1.5	1.1
Urban green spaces, waters, wetlands	2.2	2.2	2.1	1.7	1.0
Agriculture	58.1	60.4	57.5	62.2	66.8

(continued)

Table 4.3 (continued)

Variable	France	Greece	Italy	Portugal	Spain
Forests	20.0	18.9	21.6	19.0	18.5
Environmental variables					
Surface of the FUAs (km^2)	552	945	942	540	1976
SDI at FUAs level	1,126	1,199	1,151	1,172	1,210
Shannon's diversity index (H') at the landscape scale	1.27	0.74	1.02	1.38	0.91
Index of species evenness (J) Pielou at landscape scale	0.58	0.34	0.46	0.63	0.41

Table 4.4 FUAs with selective soil consumption: the cities marked in gray have an SDI class lower than the SDI at the landscape level (which indicates that a certain type of urbanization has consumed soils with higher quality than that of the overall area of the FUA)

Nation	City	Dense urban fabric	Intermediate urban fabric	Discontinuous urban fabric	Service settlements	Transportation infrastructures
FR	Aix En Provence					
	Ajaccio					
	Marseilles					
	Montpellier					
	Nice					
	Toulon					
IT	Ancona					
	Bari					
	Bologna					
	Brescia					
	Cagliari					
	Campobasso					
	Caserta					
	Catania					
	Catanzaro					
	Cremona					
	Florence					
	Foggia					
	Genoa					
	L'Aquila					

(continued)

Table 4.4 (continued)

Nation	City	Dense urban fabric	Intermediate urban fabric	Discontinuous urban fabric	Service settlements	Transportation infrastructures
	Milan					
	Modena					
	Naples					
	Padua					
	Palermo					
	Perugia					
	Pescara					
	Potenza					
	Reggio di Calabria					
	Rome					
	Salerno					
	Sassari					
	Taranto					
	Turin					
	Trento					
	Trieste					
	Venice					
	Verona					
PT	Aveiro					
	Braga					

(continued)

Table 4.4 (continued)

Nation	City	Dense urban fabric	Intermediate urban fabric	Discontinuous urban fabric	Service settlements	Transportation infrastructures
	Coimbra					
	Faro					
	Lisbon					
	Porto					
	Setubal					
SP	Alicante					
	Badajoz					
	Barcelona					
	Bilbao					
	Cordoba					
	Gijon					
	Logroño					
	Madrid					
	Malaga					
	Murcia					
	Oviedo					
	Palma de Mallorca					
	Pamplona					
	Santander					
	S. de Compostela					

(continued)

Table 4.4 (continued)

Nation	City	Dense urban fabric	Intermediate urban fabric	Discontinuous urban fabric	Service settlements	Transportation infrastructures
	Zaragoza					
	Seville					
	Toledo					
	Valencia					
	Valladolid					
	Vigo					
	Vitoria -Gasteiz					
GR	Athens					
	Kalamata					
	Candia					
	Giannina					
	Kavala					
	Larissa					
	Patras					
	Thessaloniki					
	Volos					

Table 4.5 FUAs with selective soil consumption: the cities marked in gray have an SDI class lower than the corresponding SDI value observed for dense residential settlements (indicating that mixed or discontinuous urbanization has consumed soils with higher quality than urbanization continuous and compact)

Nation	City	Intermediate urban fabric	Discontinuous urban fabric	Nation	City	Intermediate urban fabric	Discontinuous urban fabric
FR	Aix En Provence			PT	Aveiro		
	Ajaccio				Braga		
	Marseilles				Coimbra		
	Montpellier				Faro		
	Nice				Lisbon		
	Toulon				Porto		
IT	Ancona				Setubal		
	Bari			SP	Alicante		
	Bologna				Badajoz		
	Brescia				Barcelona		
	Cagliari				Bilbao		
	Campobasso				Cordoba		
	Caserta				Gijon		
It	Catania				Logrono		
	Catanzaro				Madrid		
	Cremona				Malaga		
	Florence				Murcia		
	Foggia				Oviedo		
	Genoa				P. de Mallorca		

(continued)

Table 4.5 (continued)

Nation	City	Intermediate urban fabric	Discontinuous urban fabric	Nation	City	Intermediate urban fabric	Discontinuous urban fabric
	L'Aquila				Pamplona		
	Milan				Santander		
	Modena				S. de Compostela		
	Naples				Zaragoza		
	Padua				Seville		
	Palermo				Toledo		
	Perugia				Valencia		
	Pescara				Valladolid		
	Potenza				Vigo		
	Reggio di Calabria				Vitoria -Gasteiz		
	Rome			GR	Athens		
	Salerno				Kalamata		
	Sassari				Candia		
	Taranto				Giannina		
	Turin				Kavala		
	Trento				Larissa		
	Trieste				Patras		
	Venice				Thessaloniki		
	Verona				Volos		

The results show that dense and semi-dense residential settlements and service settlements (possibly mixed with green urban spaces) represent more sustainable urban morphologies than discontinuous ones, preserving the quality of the surrounding agricultural soils and the basic environmental services they provide (Cecchini et al. 2019). The combined effects of degradation processes (e.g. soil sealing, contamination and compaction), microclimatic conditions (influenced by urban heat), scarce and fragmented vegetation cover and increasing anthropogenic pressure contribute to deteriorating the environmental conditions of peri-urban areas and reducing their socio-economic value (Kairis et al. 2013; Kelly et al. 2015; Kosmas et al. 2016).

Selective sprawl-driven land use of high quality should be more carefully considered in urban planning. It becomes increasingly necessary for planning to identify places without current use and urban gaps intended for construction (Paül and Tonts 2005). In this sense, agricultural areas with intermediate and poor land quality surrounding cities can be partly destined for forms of urban land-saving expansion (Biasi et al. 2015b; Duvernoy et al. 2018; Perrin et al. 2018), in light of sustainable management of peri-urban lands (Salvati et al. 2018a). At the same time, agricultural areas with a high land quality should be more carefully preserved from irresponsible urban expansion toward current use (e.g. agricultural or forestry). Especially open areas with high land quality near peri-urban forests and agricultural areas, seen as potential sites for urban development, would require targeted conservation strategies (Kairis et al. 2015).

References

Aguilera F, Valenzuela LM, Botequilha-Leitão A (2011) Landscape metrics in the analysis of urban land use patterns: a case study in a Spanish metropolitan area. Landscape and Urban Plann 99:226–238. https://doi.org/10.1016/j.landurbplan.2010.10.004

Bajocco S, Ceccarelli T, Smiraglia D, Salvati L, Ricotta C (2016) Modeling the ecological niche of long-term land use changes: the role of biophysical factors. Ecol Ind 60:231–236. https://doi.org/10.1016/j.ecolind.2015.06.034

Bajocco S, De Angelis A, Salvati L (2012) A satellite-based green index as a proxy for vegetation cover quality in a Mediterranean region. Ecol Indic 23:578–587. https://doi.org/10.1016/j.ecolind.2012.05.013

Bajocco S, Salvati L, Ricotta C (2011) Land degradation versus fire: a spiral process? Prog Phys Geogr: Earth Environ 35:3–18. https://doi.org/10.1177/0309133310380768

Barbero-Sierra C, Marques MJ, Ruíz-Pérez M (2013) The case of urban sprawl in Spain as an active and irreversible driving force for desertification. J Arid Environ 90:95–102. https://doi.org/10.1016/j.jaridenv.2012.10.014

Basso F, Bove E, Dumontet S, Ferrara A, Pisante M, Quaranta G, Taberner M (2000) Evaluating environmental sensitivity at the basin scale through the use of geographic information systems and remotely sensed data: an example covering the Agri basin (Southern Italy). CATENA 40:19–35. https://doi.org/10.1016/S0341-8162(99)00062-4

Biasi R, Brunori E, Smiraglia D, Salvati L (2015a) Linking traditional tree-crop landscapes and agro-biodiversity in central Italy using a database of typical and traditional products: a multiple risk assessment through a data mining analysis. Biodiver Conserv 24:3009–3031. https://doi.org/10.1007/s10531-015-0994-5

Biasi R, Colantoni A, Ferrara C, Ranalli F, Salvati L (2015b) In-between sprawl and fires: long-term forest expansion and settlement dynamics at the wildland–urban interface in Rome, Italy. Int J Sustain Dev World Ecol 22:467–475. https://doi.org/10.1080/13504509.2015.1064488

Cecchini M, Zambon I, Pontrandolfi A, Turco R, Colantoni A, Mavrakis A, Salvati L (2019) Urban sprawl and the 'olive' landscape: sustainable land management for 'crisis' cities. GeoJournal 84:237–255. https://doi.org/10.1007/s10708-018-9848-5

Ciommi M, Chelli FM, Salvati L (2019) Integrating parametric and non-parametric multivariate analysis of urban growth and commuting patterns in a European metropolitan area. Qual Quan 53:957–979. https://doi.org/10.1007/s11135-018-0798-2

Coisnon T, Oueslati W, Salanié J (2014) Urban sprawl occurrence under spatially varying agricultural amenities. Regl Sci Urban Econ 44:38–49. https://doi.org/10.1016/j.regsciurbeco.2013.11.001

Colantoni A, Zambon I, Gras M, Mosconi EM, Stefanoni A, Salvati L (2018) Clustering or scattering? the spatial distribution of cropland in a Metropolitan region, 1960–2010. Sustainability 10:2584. https://doi.org/10.3390/su10072584

Cuadrado-Ciuraneta S, Durà-Guimerà A, Salvati L (2017) Not only tourism: unravelling suburbanization, second-home expansion and "rural" sprawl in Catalonia, Spain. Urban Geogr 38:66–89. https://doi.org/10.1080/02723638.2015.1113806

Di Feliciantonio C, Salvati L, Sarantakou E, Rontos K (2018) Class diversification, economic growth and urban sprawl: evidences from a pre-crisis European city. Qual Quan 52(4):1501–1522

Duvernoy I, Zambon I, Sateriano A, Salvati L (2018) Pictures from the other side of the fringe: urban growth and peri-urban agriculture in a post-industrial city (Toulouse, France). J Rural Stud 57:25–35

Ferrara A, Salvati L, Sabbi A, Colantoni A (2014a) Soil resources, land cover changes and rural areas: towards a spatial mismatch? Sci Total Environ 478:116–122. https://doi.org/10.1016/j.scitotenv.2014.01.040

Ferrara C, Salvati L, Tombolini I (2014b) An integrated evaluation of soil resource depletion from diachronic settlement maps and soil cartography in peri-urban Rome, Italy. Geoderma 232–234:394–405. https://doi.org/10.1016/j.geoderma.2014.05.020

Ferrari M et al. 2003. Ecologia agraria. AbeBooks. Edagricole Scolastico, Milano, Italy

Kairis O, Karavitis C, Kounalaki A, Salvati L, Kosmas C (2013) The effect of land management practices on soil erosion and land desertification in an olive grove. Soil Use Manage 29:597–606. https://doi.org/10.1111/sum.12074

Kairis O, Karavitis C, Salvati L, Kounalaki A, Kosmas K (2015) Exploring the impact of overgrazing on soil erosion and land degradation in a dry Mediterranean agro-forest landscape (Crete, Greece). Arid Land Res Manage 29:360–374. https://doi.org/10.1080/15324982.2014.968691

Karamesouti M, Detsis V, Kounalaki A, Vasiliou P, Salvati L, Kosmas C (2015) Land-use and land degradation processes affecting soil resources: evidence from a traditional Mediterranean cropland (Greece). CATENA 132:45–55. https://doi.org/10.1016/j.catena.2015.04.010

Kelly C, Ferrara A, Wilson GA, Ripullone F, Nolè A, Harmer N, Salvati L (2015) Community resilience and land degradation in forest and shrubland socio-ecological systems: evidence from Gorgoglione, Basilicata, Italy. Land Use Policy 46:11–20. https://doi.org/10.1016/j.landusepol.2015.01.026

Kosmas C, Karamesouti M, Kounalaki K, Detsis V, Vassiliou P, Salvati L (2016) Land degradation and long-term changes in agro-pastoral systems: an empirical analysis of ecological resilience in Asteroussia—Crete (Greece). CATENA 147:196–204. https://doi.org/10.1016/j.catena.2016.07.018

Lamonica GR, Recchioni MC, Chelli FM, Salvati L (2020) The efficiency of the cross-entropy method when estimating the technical coefficients of input–output tables. Spat EconAnal 15(1):62–91

Lafortezza R, Carrus G, Sanesi G, Davies C (2009) Benefits and well-being perceived by people visiting green spaces in periods of heat stress. Urban for Urban Green 8:97–108. https://doi.org/10.1016/j.ufug.2009.02.003

Montanarella L (2007). Trends in land degradation in Europe. In: Mannava V, Sivakumar K, Ndiang'ui N (eds) Climate and land degradation. Environmental science and engineering. Springer, Berlin, Heidelberg, pp 83–104. https://doi.org/10.1007/978-3-540-72438-4_5

Paül V, Tonts M (2005) Containing urban sprawl: trends in land use and spatial planning in the Metropolitan region of Barcelona. J Environ Plan Manage 48:7–35. https://doi.org/10.1080/0964056042000308139

Perrin C, Nougarèdes B, Sini L, Branduini P, Salvati L (2018) Governance changes in peri-urban farmland protection following decentralisation: a comparison between Montpellier (France) and Rome (Italy). Land Use Policy 70:535–546

Portnov BA, Safriel UN (2004) Combating desertification in the Negev: dryland agriculture vs. dryland urbanization. J Arid Environ 56:659–680. https://doi.org/10.1016/S0140-1963(03)00087-9

Recanatesi F, Clemente M, Grigoriadis E, Ranalli F, Zitti M, Salvati L (2016) A fifty-year sustainability assessment of Italian agro-forest districts. Sustainability 8:32. https://doi.org/10.3390/su8010032

Salvati L (2014a) A socioeconomic profile of vulnerable land to desertification in Italy. Sci Total Environ 466–467:287–299. https://doi.org/10.1016/j.scitotenv.2013.06.091

Salvati L (2014b) Agro-forest landscape and the 'fringe' city: a multivariate assessment of land-use changes in a sprawling region and implications for planning. Sci Total Environ 490:715–723. https://doi.org/10.1016/j.scitotenv.2014.05.080

Salvati L (2014c) Towards a polycentric region? The socio-economic trajectory of Rome, an 'Eternally Mediterranean' city. Tijdschr Econ Soc Geogr 105:268–284

Salvati L, Carlucci M, Grigoriadis E, Chelli FM (2018a) Uneven dispersion or adaptive polycentrism? Urban expansion, population dynamics and employment growth in an 'ordinary' city. Review Reg Res 38:1–25. https://doi.org/10.1007/s10037-017-0115-x

Salvati L, Ferrara A, Chelli F (2018b) Long-term growth and metropolitan spatial structures: an analysis of factors influencing urban patch size under different economic cycles. Geogr Tidsskrift-Dan J Geogr 118:56–71. https://doi.org/10.1080/00167223.2017.1386582

Salvati L, Gemmiti R, Perini L (2012) Land degradation in Mediterranean urban areas: an unexplored link with planning? Area 44:317–325. https://doi.org/10.1111/j.1475-4762.2012.01083.x

Salvati L, Serra P (2016) Estimating rapidity of change in complex urban systems: a multidimensional, local-scale approach. Geogr Anal 48:132–156. https://doi.org/10.1111/gean.12093

Salvati L, Tombolini I, Ippolito A, Carlucci M (2018c) Land quality and the city: monitoring urban growth and land take in 76 Southern European metropolitan areas. Environ Plann B: Urban Anal City Sci 45:691–712. https://doi.org/10.1177/0265813516684827

Salvati L, Zambon I, Chelli FM, Serra P (2018d) Do spatial patterns of urbanization and land consumption reflect different socioeconomic contexts in Europe? Sci Total Environ 625:722–730. https://doi.org/10.1016/j.scitotenv.2017.12.341

Santos M, Cabral JA (2004) Development of a stochastic dynamic model for ecological indicators' prediction in changed Mediterranean agroecosystems of north-eastern Portugal. Ecol Indic 3:285–303. https://doi.org/10.1016/j.ecolind.2003.11.007

Serra P, Vera A, Tulla AF, Salvati L (2014) Beyond urban–rural dichotomy: exploring socioeconomic and land-use processes of change in Spain (1991–2011). Appl Geogr 55:71–81. https://doi.org/10.1016/j.apgeog.2014.09.005

Wilson GA, Juntti M (ed) (2005) Unravelling desertification. Wageningen Academic Publishers, Wageningen, The Netherlands. https://doi.org/10.3920/978-90-8686-559-8

Zambon I, Benedetti A, Ferrara C, Salvati L (2018) Soil matters? a multivariate analysis of socioeconomic constraints to urban expansion in Mediterranean Europe. Ecol Econ 146:173–183. https://doi.org/10.1016/j.ecolecon.2017.10.015

Zambon I, Serra P, Sauri D, Carlucci M, Salvati L (2017) Beyond the 'Mediterranean city': socioeconomic disparities and urban sprawl in three Southern European cities. Geografiska Annaler: Series B, Human Geography 99:319–337. https://doi.org/10.1080/04353684.2017.1294857

Chapter 5
Preserving Land Quality in European Metropolis

Abstract The initiatives of the Member States of the European Union aimed at soil protection vary and focus on the degradation processes that each of them considers as priorities. Mediterranean countries have adopted national action programs to combat land degradation, to which they are particularly sensitive both for climatic conditions (drought, water scarcity, intense rainfall) and for anthropic activities (such as deforestation). In their programs, they identify the actions necessary to address the problem of land degradation and which, therefore, are implicitly in favor of land quality. In this chapter, we discuss some management and governance aspects considering the study cases of Italy, Spain, Portugal, Greece or France. In addition, we propose a wide range of cultural and good practices, which are nowadays implemented by some countries or should be taken into account.

Keywords Land quality · Governance · Metropolis · Good practices · Cultural aspects

5.1 Management and Governance Aspects

The initiatives of the Member States of the European Union aimed at soil protection vary and focus on the degradation processes that each of them considers as priorities. While efforts in central and northern Europe focus on the problem of pollution and soil sealing, initiatives have been launched in the Mediterranean countries for the management of erosion and land degradation under the United Nations Convention for the Fight against Drought and Desertification (Prokop et al. 2018).

Mediterranean countries have adopted national action programs to combat land degradation, to which they are particularly sensitive both for climatic conditions (drought, water scarcity, intense rainfall) and for anthropic activities (such as deforestation). In their programs, they identify the actions necessary to address the problem of land degradation and which, therefore, are implicitly in favor of land quality (Salvati, Zambon, et al. 2018). In particular:

– National Action Program of Greece: (i) identify the extent of areas at risk of desertification; (ii) promote the sustainable use of soil and water; and, (iii) the social and economic rehabilitation of the affected areas from land degradation.

I. Tombolini et al., *Land Quality and Sustainable Urban Forms*, Springer Geography,
https://doi.org/10.1007/978-3-030-94732-3_5

- Italian program: above all at reducing the risk of floods and landslides. It focuses on water regulation and coordination of sectoral policies that influence the cycle. High-risk areas for floods and landslides are defined in the document.
- Spanish program, which certifies that about a third of the nation is already seriously threatened by desertification, foresees a series of actions to manage water resources sustainably and to prevent forest fires.
- Portuguese program focuses on measures for the recovery of degraded areas, spread all along the coast, encouraging the population to move to areas with less population density.

Despite these general lines of action, it is possible to affirm that at the European level there is no strategic vision for soil protection that adequately takes into account its functions. Furthermore, there are no binding targets that specifically concern soil sealing and soil consumption issues, which, as seen in the first chapters, also pose a serious threat to land degradation and affect the maintenance of good land quality (Recanatesi et al. 2016). Soil sealing is explicitly considered only by non-binding guidelines and is partially related to the flood directive (Munafò et al. 2013).

Quantitative land-use limits currently exist in Germany (where limits are defined in hectares for days, programmed from year to year), Belgium, Luxembourg and The Netherlands (limits are based on internal urban development) or England (limits are based on the development of brownfield sites). The definition of similar targets was not found in the urban policies of the nations where the metropolises object of the analysis of this research are located (Salvati 2014b). However, as detailed below, Greece, Italy, Spain and Portugal have also started a journey in this direction.

5.1.1 Greece

The intensity of land use is very high in Greece because very dense settlements are widespread in the country. The percentage of the urban population is over 60 and 41% of the population lives in cities with more than 100,000 inhabitants. According to Eurostat data, 3.5% of the territory is waterproofed, a rather moderate percentage compared to the other Member States of the European Union. The regions where the soil is subjected to strong pressure are certainly the agglomeration of Athens and the coastal areas, especially for tourism reasons.

The basic rules for the protection of the natural and cultural environment as well as for regional and urban planning are included in article 24 of the Greek Constitution. The Constitution places environmental protection and spatial planning under the responsibility of the state and specifically provides for the protection of forests, the safeguarding of cultural heritage and the contribution of landowners for the safety of lands for social structures and services.

However, there are still no specific actions dedicated to soil protection, or that foresee actions for the control of soil sealing and soil consumption. Some aspects of soil protection are fragmented in national legislation. It is worth mentioning:

- Law 1650/1986 (Official Journal A 160 / 16.10.1986) on the protection of the environment;
- Law 16,190/1335/1997 "measures and conditions for the protection of waters from nitrates from agricultural sources";
- Law 3199/2003 (Official Journal A 280 / 09.12.2003) on the protection and management of water;
- National planning (decision 6876/4871/2008) and special measures for industry (judgment 11,508/2009) and tourism (decision 24,208/2009);
- Ministerial decision 36,259/1757 / E103 / 2010 "measures, modalities and program for the alternative management of excavation, construction and demolition waste";
- Law 168,040/2010 (Official Journal B 1528 / 07.09.2010) for the "determination of the criteria for the classification of quality agricultural land";
- Law 3937/2011 (Official Journal A 60 / 31.03.2011) on the conservation of biodiversity;
- Law 4042/2012 (Official Journal A 24 / 13.02.2012) on the criminal protection of the environment;
- Law 36,060/1155 / E.103 / 2013 to set the "guidelines, measures and procedures for the prevention and control of pollution from industrial activities, in compliance with the provisions of Directive 2010/75 / EU of the Parliament European Council and the Council of 24 November 2010" .

A draft law for the protection and sustainable use of the soil was also prepared in April 2014, to establish a global framework for the protection and sustainable use of the soil. It was based on the guidelines guidance provided by the decisions of international organizations (such as the UN) and, above all, based on the draft European directives, detailed in the discussion period 2007–2009 by the Council on the Environment of the European Union but not yet institutionalized. The proposal promotes the integration of soil protection into other environmental policies, the conservation of soil functions in the context of sustainable use, the prevention and mitigation of the effects of threats to soil and the restoration of degraded soils. In particular, article 4 concerns the containment of soil consumption and soil sealing.

In paragraph 1 of the article, it is specified that the ministry must adopt appropriate measures to limit the waterproofing of the soil, for example with a necessary and planned residential development, thus, reducing the contraction of rural areas and the restoration of abandoned polluted sites.

5.1.2 Italy

Unlike most European countries, Italy does not have a national-territorial development plan, for the coordination of regional and local plans. The 20 Italian regions have a high degree of autonomy about territorial planning, and the most important planning document is the Regional Territorial Plan (PTR), which concerns the regulation relating to land use, the development of the territory on a large scale and planning of infrastructures such as the road network and railways. The regional territorial plan is drawn up jointly by representatives of the provinces, municipalities, private entities and other interested parties (Clemente et al. 2018).

As above-anticipated, the regulatory and binding tools developed for the protection of the Italian territory have proven to be ineffective in combating the phenomenon of soil consumption, and indeed, in many cases, have supported it. The landscape and environmental constraint, a very useful tool for identifying and preserving emerging territorial goods and values, is not a tool for regulating land consumption; on the contrary, it produces what can be defined as the "margin effect", generating "shorelines" around the protected areas, behind which, paradoxically, the proximity income thickens. The affixing of constraints on rural areas or of recognized landscape value, when preserving them, makes the locations on the margins (in particular the residence) attractive, which enjoy the institutional guarantee that that endowment of greenery and landscape value will be available to an indefinite time. The result is the encirclement of the protected area, clearly visible in the case of regional parks of an older constitution (Di Simine et al. 2013).

After highlighting that, despite the compendious body of environmental protection rules of our country, there are no regulatory tools and prevention of the phenomenon of soil consumption. It is worth mentioning the first bill for the limitation of soil consumption, which dates back to 2012, when the then Minister of Agricultural, Food and Forestry Policies presented the Report "Building the future: defending agriculture from cementing" and the bill "valorising agricultural areas and containing land consumption", not approved due to the early end of the Legislature.

A new government initiative bill was presented in 2014 and, after more than two years of discussion, approved in the Chamber on 12 May 2016. Also, based on the data contained in the latest ISPRA reports on Soil Consumption and some considerations related to recognized limits of the law, the Commissions meeting the Territory and Environment and Agriculture of the Senate, between 2016 and 2017, following a thorough cycle of hearings, arrived at the significant revision of some articles of the text of the law and the introduction of important innovative elements capable of making the standard more effective. This focused on a particular reference to the system of definitions, the identification, implementation and monitoring (Lamonica et al. 2020) of progressive limits to soil consumption, reuse and urban regeneration and the protection of green areas in an urban area. In particular, the text envisaged a progressive reduction in land use of at least 15% every three years. However, even in this case, the end of the legislature did not allow for final approval.

At the beginning of this legislature, some bills were presented which, in part, take over and update the previous text and which hopefully will soon guarantee our country a fundamental law for the protection of the environment, the ecosystem and of the Italian landscape (letter s of the 2nd paragraph of article 117 of the Constitution). The Agriculture and Environment commissions of the Senate have meanwhile started the joint examination of two bills on the subject and are working to reach a unified text.

These are bills 86 (23 March 2018) and 164 (27 March 2018). Legislative Decree 86 "Provisions for the reduction of land consumption as well as a delegation to the

Government regarding the regeneration of degraded urban areas" was presented by Senator Loredana De Petris (Mixed Group) and defines soil as "common good and non-renewable resource functions and produces ecosystem services. Also according to the prevention and mitigation of hydrogeological instability events, mitigation strategies and adaptation to climate change, the reduction of phenomena that cause erosion, loss of organic matter and biodiversity".

Bill 164 "Provisions for stopping land use, reuse of built land and for the protection of the landscape", presented by Senator Paola Nugnes (M5S), incorporates the contents of the text elaborated by the Forum "Salviamo il Paesaggio" and made available to political forces. According to bill 164, the soil plays a "fundamental role for the survival of living beings", which is why the indifference of the actions aimed at preserving it from further possible transformations, from erosion and cementification is highlighted. Although both bills pursue the goal of stopping soil consumption, bill 86 provides for systematically, while bill 164 provides for immediate measures. In addition, Legislative Decree 86 provides for zero land consumption by 2050, consistent with the target set by the European Union. It asks the regions to define the progressive reduction of land consumption, which must be at least 20% every three years compared to the land consumption recorded in the previous three years, both for permanent and reversible consumption. The draft also provides for the establishment of a green belt around the inhabited centers and the obligation for the Municipalities to survey the abandoned and unused or abandoned buildings and areas to check if they can be subject to a regeneration program capable of avoiding the consumption of new soil.

Legislative Decree 164, on the other hand, sanctions the immediate halt of land consumption and the modification of urban planning tools in the Municipalities aimed at eliminating the construction forecasts involving land consumption in agricultural, natural and semi-natural areas. To achieve these objectives, the bill introduces a series of obligations to be paid by the Municipalities, to be adopted within six months from the date of entry into force of the regulation. The obligations consist of: (i) identifying the areas or buildings to be subjected primarily to reuse and urban regeneration; (ii) in the preparation of a plan that identifies and delimits the existing urbanized area; and, (iii) in the execution of a municipal building census that allows the creation of a database of public and private building heritage to be recovered.

In the absence of a national standard, the framework of the regional legislation is rather heterogeneous, including provisions, regulations or principles included in laws aimed at limiting soil consumption and urban regeneration. Many regions have specific rules on land use; others have set or set targets on the subject under local government laws. In some regions, the principle of containment of land use is included in rules relating to urban redevelopment or regeneration, often understood as an alternative to new land use. However, practically everywhere, the definition of land consumption is not consistent with the European and national one or, in any case, there are significant derogations or exceptions relating to types of interventions and transformations of the territory that are not included in the calculation (and therefore in the limitation) but which are a clear cause of land use. Among these, there are frequently, for example, interventions envisaged by current municipal urban plans, public works of supra-municipal interest, construction or expansion of production sites, rural buildings, infrastructures or public services and urban densification interventions (Salvati 2014a).

5.1.3 France

In France, land use and soil sealing mainly affect metropolitan areas and coastal regions. According to Eurostat data, 5.2% of the territory is waterproofed. The fragmentation of the landscapes and the growing distances of the commuters are becoming increasingly important in French environmental policy (Duvernoy et al. 2018).

In 2006, the National Strategy for Sustainable Development underwent a revision process and a new sustainability objective was defined concerning the reduction of soil consumption, entitled "stop the disproportionate growth of the artificial surface compared to the population growth building new infrastructures on already artificial lands". The strategy for sustainable development has been incorporated into the "Grenelle law", an expression that is used to refer to two laws of the Grenelle Environment Forum, which have made important changes to environmental legislation and have placed greater emphasis on reducing land use:

- Grenelle I law (law n. 2009–967 of 3 August 2009), relating to the implementation of the 268 commitments of the Grenelle Environnement;
- Grenelle II law (law of 12 July 2010), on the national commitment to the environment, in which the more specific objectives and provisions are declined, in 57 articles grouped in 6 titles, to formulate a framework of action on the ecological emergency.

The law aims to establish a complete legal framework for the protection of the environment, the reduction of energy consumption, the improvement of economic and social stability. The policy framework is based on six main action lines, each of which is supported by legal requirements, pilot applications and research. The most relevant line of action for the reduction of soil consumption and its waterproofing is "the improvement of the energy standards of buildings and the harmonization of territorial planning", which provides for energy-efficient urban structures, supporting internal urban development and avoiding further soil consumption (Perrin et al. 2018).

With Grenelle II, the PLU (Plan local d'urbanisme, the main urban planning tool at municipal or inter-municipal level) becomes one of the new tools aimed at ensuring economic management of the soil. The law assigns public actors the mission of ensuring effective control of the consumption of the natural, agricultural and forestry area. However, there are doubts about the fact that the assignment to the PLU of this objective is truly effective for the reduction of land consumption. The concern is that local actors can open natural areas to urbanization, as an economic and profitable solution, seeing the price of their land multiply, rather than opting for densification, being a more complicated, expensive and legally complex operation in urban areas.

5.1.4 Spain

Compared to the other EU Member States, the percentage of artificial surfaces and sealed soils is still not high (3.5% according to Eurostat). The negative impacts that derive from urban sprawl and soil sealing are limited to a few hotspots, which are large urban areas and coastal areas.

In Spain, the coast is particularly affected by land use; in fact, 45% of the population lives in coastal municipalities which represent only 7% of the territory (Prokop et al. 2018). Awareness of the negative effects of urban sprawl and land use is increasing and the first measures have already been started. These include the protection of coastal areas from further land use, the strengthening of the redevelopment of brownfield sites and the regulation of the real estate market (Cuadrado-Ciuraneta et al. 2017).

Spatial planning is the responsibility of the 17 Regions (Autonomous Communities) and each region has its detailed legislation governing planning. The central document is the planning law (Ley de Suelo), which regulates the right to build and the value of the land. The first soil law was enacted in 1956 and the most recent in 2007. Following a legislative reform that took place in 2015, the current law was merged with the law on urban rehabilitation, right of regeneration and renovation, with the name of Ley del Suelo y rehabilitación Urbano. The planning law defines the general principles of spatial planning, The actual implementation is under the responsibility of the autonomous communities (Salvati and Carlucci 2016).

The most important documents for spatial planning are the territorial general plans (PTG, Plan Territorial General) of the 17 autonomous communities, which are continually reviewed. The awareness of the massive pressures on land use, which occurs mainly in coastal regions, is increasingly considered in these documents. To give an example, the Catalonia planning law promotes the protection of natural and agricultural areas, the reduction of land consumption and the compactness of urban forms. Coastal protection is a constitutional duty of the national and regional government. In Spain, the main legislative act in this regard is the law 22/1988 (28 July 1988). This governs land cover in coastal areas, defining a protection belt, consisting of a 100-m wide belt in which construction is prohibited and a 500-m belt within which the building is strictly controlled. Spain is also developing a strategy for sustainable development in coastal areas (Estrategia para la Sostenibilidad de la Costa). One of the measures contained in the strategy is the definition of a public dividing line to guarantee public access and use, regulate the rational use of its assets and ensuring adequate quality of coastal waters. The protection of the coast and, in particular, the prohibitions of new settlement development, are seriously considered by the Spanish authorities. Violations of this prohibition and the obligation to demolish buildings built within the protected belt are regularly reported in the Spanish press.

5.1.5 Portugal

Transport infrastructures in Portugal are among the densest in Europe, with the highest number of kilometers of roads per number of inhabitants. Between 1990 and 2006, the country faced enormous urban development and the amount of artificial surface area almost doubled (Prokop et al. 2018). This significant urban expansion has mainly affected the coastal areas and urban agglomerations of Lisbon, Setubal and Porto, but currently, the suburbs and second homes in the tourist regions are the main consumers of soil. Uncontrolled urban growth, ruined urban centers and the contraction of rural areas are of particular concern. These critical issues are considered in the three reference documents for spatial planning.

The main document is the National Strategy for Sustainable Development (NSDS), approved in August 2006. The document refers specifically to the intention to reverse the trend of extensive and low-quality urban growth and encourage urban redevelopment and the recovery of degraded areas, promoting a higher quality of life standards.

The National Spatial Planning Program (NPSP), approved by Parliament in September 2007, establishes the main options that are relevant for the organization of the country, in line with NSDS and with the values contained in the concept of regeneration. At the regional level, this program is implemented through regional spatial planning plans and, at the municipal level, from municipal general plans. Specific objectives are to requalify urban areas, preserve the available natural resources and better coordinate urban growth (Salvati 2013).

Polis XXI, approved in March 2007, is the urban policy program for the sustainable development and national cohesion of Portuguese cities. It consists of a series of integrated urban policy tools aimed at promoting urban regeneration, competitiveness and innovation through networking and improving the quality of life and the environment in cities. It highlights urban regeneration as an essential dimension of the cohesion of cities, crucial for the quality of life.

Portugal has two levels of territorial governance: the national and municipal levels (308 municipalities). In addition, there are two autonomous regions (the Azores and Madeira islands). The national government has four distinct functions related to land-use policies. First, it provides the legal framework that governs planning at the national, regional and local level; secondly, it defines national strategies and sectoral policies aimed at integrated, cohesive and sustainable territorial development of the country; thirdly, it assigns national and European Union funds to specific territories and projects; fourthly, it provides technical assistance at the regional level and for municipal planning.

In addition to their direct responsibility for land-use planning, municipalities also influence land use concerning the construction of public buildings and municipal infrastructure. Several other public authorities and companies influence land-use policies in Portugal, most of which are controlled by the national government. Among these is the Institute for Nature Conservation and Forestry, which is responsible for ensuring that spatial planning follows the principles of sustainable development.

In 2014–2015, an important reform of the land-use system in Portugal took place, to strengthen the strategic dimension of the planning process. It has led to a clearer division between national and regional programs and has created the possibility for municipalities to form inter-municipal entities for joint planning and for changing land-use categories in an attempt to contain urban sprawl. The results and methodology of the exploratory analysis illustrated in the previous chapter could be exploited in the monitoring programs and policies, under development, in favor of sustainable soil management. An understanding of how different forms of urban expansion impact land quality can indeed constitute a relevant information base for a reasoned assessment of the consumption of natural resources caused by urbanization. In tackling new urban issues, it could be equally important to start from a reflection of urban plans and how they could solve, also through integration processes, the "sum of absences" that at various levels characterizes the most peripheral urban areas: the absence of services and collective spaces, but also the absence of functions and uses of the land.

The European Commission (EC) itself, in 2012, published a report[1] on the most effective mechanisms to limit, mitigate and compensate for soil sealing (or "soil waterproofing") which, as has already been widely discussed, is one of the anthropogenic phenomena that has the greatest impact on land degradation in metropolitan landscapes. From the measures proposed by the European Commission, it emerges how important the role of planning is in achieving more sustainable land use. This calls for the inclusion in the spatial planning of mechanisms that take into account the quality, characteristics and functions of the soil, which counterbalance individualistic objectives and interests. In this regard, Glæsner et al. (2014) highlight how effective spatial planning, capable of protecting even the non-economic functions of the soil, would guarantee the maintenance of all its functions; likewise, Tobias (2013) believes that spatial planning can compensate for soil sealing, helping to preserve ecosystem services.

Strategic spatial planning can also be an effective approach to stop and even reverse the processes that lead to land degradation and that can improve land quality, in particular those deriving from urbanization and the consequent waterproofing of soils. It represents a development of vision, which over time can be modeled based on contextual socio-political and institutional complexity, involving various actors and various sectors (Oliveira and Hersperger 2018). Another advantage of this

[1] European Commission (2012), *Guidelines on Best Practice to Limit, Mitigate or Compensate Soil Sealing.*

approach is that it allows better management of certain issues for a longer period than traditional spatial planning, which offers rather short-term prospects. Furthermore, strategic spatial planning processes are often non-binding and therefore less tied to spatial planning and legally binding political instruments, thus offering the possibility to experience further benefits that allow advancing scientific frontiers and political agendas. The lack of constraints, however, can decrease the effectiveness of the measures that protect the soil.

Citing Oliveira et al. (2018), in the literature, few significant examples can be identified of how through strategic spatial planning we have tried to pursue specific objectives:

- support for sustainable agriculture development (Jurgens 1993);
- balance of urban and rural development prevention against environmental degradation (Healey 2004);
- protection of agricultural land (therefore fertile) from uncontrolled urban sprawl (Scott 2006; Satterthwaite et al. 2010).

As the same Oliveira et al. (2018) suggest strategic spatial planning should be more practice-oriented in terms of actions against land degradation, using specific data on land use and quality. It should also be further investigated how to effectively integrate policies aimed at controlling urban sprawl into strategic spatial planning. Existing research reveals that establishing stringent restrictions and heavy financial penalties on urban developments outside the planning limit can contain urban sprawl. However, doubts remain about whether this type of solution is truly effective in the current context characterized by rapid urbanization and increasing pressure from private economic actors on the consumption of fertile land, as well as green spaces in general. In this context, research on strategic planning could benefit from deepening the interrelation with that on socio-ecological systems, to better understand how land degradation affects people and the environment, due to its effects on agricultural land, on food production, natural habitats and recreational green areas (Salvati, Ferrara, et al. 2018).

One of the most used indicators for the assessment of socio-ecological systems is the per capita forest area (AFP), being positively correlated with the health status of the ecosystem (Li and Pan 2014). Tomao et al. (2017) proposed an analysis through this indicator to evaluate soil management in a region of southern Europe, Attica (Greece), in the decades between 1960 and 2010. This region, as seen previously, is considered an emblematic case of the urban transformation that took place in the metropolitan regions of the northern Mediterranean basin, where the dispersive and polycentric development of the settlements altered the typical compact spatial organization. In 1960, the highest AFP values were observed in the municipalities located in the northern part of Attica, which represent an example of an unplanned "green belt" around the consolidated settlement area of Athens. The decrease in the forest area has caused a narrowing of this belt, especially after 1990, also leading to the loss of ecological connectivity between the natural areas surrounding Athens.

The per capita forest area decreased significantly in the 1980s and 1990s in the fringe areas surrounding Athens and in the districts subject to urban sprawl (such as Messoghia). The important changes observed both at regional and local level underline a polarization between areas with low efficiency in land use and areas with high efficiency along the urban–rural gradient. As the authors state, this phenomenon can be explained through the increase in forest cover and, in particular, with the so-called "forest transition theory", which refers to the recovery of forest surfaces after a long period of decline, which follows the abandonment of agricultural activity. In Attica, few afforestation or reforestation interventions have been planned in recent decades.

On the one hand, this increase in forest cover has a positive effect because the supply of ecosystem services increases; on the other, these uncontrolled and unplanned changes in the landscape can pose a serious threat to the extension of the boundaries of the wildland-urban interface (WUI) (Barbati et al. 2013; Chas-Amil et al. 2013; Biasi et al. 2015). The biomass of newly formed forests, for example, is more susceptible to the risk of fire (Bajocco et al. 2011, 2015; Elia et al. 2014). In such a perspective, forest planning (through the recovery of agriculture or forest management plans) in urban and peri-urban areas can contribute both to landscape integration, through the reconnection of natural areas (Corona et al. 2002), and the enhancement of soil functions, recovering its ecological functionality. Improving the functionality, quality and accessibility of new forests should become a management priority (Bajocco et al. 2012).

5.2 Cultural Aspects and Good Practices

As mentioned in the previous chapters, the soil, for its important intrinsic and extrinsic qualities, and for the flows of energy and matter that it continuously exchanges with the surrounding environment, should be considered a living system that interacts with the landscape, absorbing its transformations. Soil should therefore play an important role in the landscape project, which must understand this fundamental resource in its thickness of support and infrastructure that supports artificial settlements, and as the affirmation of the deployment of the material and immaterial networks that make up the landscape. In this context, for the intervention planning strategies, the theme of dross capes—or scraps of waste—by Alan Berger can be of inspiration.

The landscapes of the waste are interstices, spaces in-between in the urban fabric of the city, they are areas that sediment in the wake of the spatial and socio-economic process of de-industrialization as empty and marginal spaces between buildings. A margin that also occurs externally, where the clear boundaries between the agricultural landscape and the urbanized landscape have turned into a frayed edge (Antrop 2004). In examining the empty (or open) spaces of cities as an opportunity for transformation, Berger (2006) investigates the etymological relationships between the words *vast*, *waste* e *dross*.[2] In particular, it highlights how dross (in Italian "waste")

[2] The Latin term vastus is the etymological root of both the term vast and waste. Vastus is an adjective that primarily has the meaning of emptiness, depopulated, desert, but also made desert by

derives from the combination of natural and anthropic processes, thus becoming a product generated by practices external to it. By transferring the reasoning to the urban space, the author affirms that the waste (dross) is considered as a natural component of every city that develops dynamically and is therefore an indicator of urban development. Recycling the waste of industrial production and reconverting an immense abandoned property is a complex and difficult process, but it is not to be overlooked.

The redevelopment of abandoned areas is a strategy that certainly contributes to preserving land quality in European metropolises, as it has the objective of avoiding new land consumption, reusing land already developed for new infrastructure projects. For example, France has more than 20 public agencies for the development of the territory (Etablissement Public Foncier), which among other activities deal with the requalification of abandoned areas for social housing. The Provence-Alpes-Côte d'Azur region also has it.

The tensions to which the region is subject due to its exceptional attraction, the natural fragility of its spaces and the imbalances affecting its territories require, in response, the mobilization of public actors in the conduct of coordinated and convergent policies in regional planning. One of the most important issues concerns the control of urban sprawl and the creation of conditions for sustainable development, containing soil consumption and preserving the integrity and quality of natural and agricultural sites. In Marseille, near an underground station, an ambitious operation was carried out, the result of the transformation of a former peripheral industrial complex, now absorbed by the city center (Fig. 5.1).

Although heavily constructed, this space was degraded and obsolete. The project made it possible to reopen it toward one's environment, recreating permeability and promoting urban functions: housing, commerce, spaces and public structures. Another rehabilitation project operated by these French agencies (Etablissement Public Foncier) concerns another coastal city, Toulon, where 103 student accommodation and 42 social housing were built in a particularly degraded area (the island of Baudin) (Fig. 5.2).

An important example of redevelopment concerns Portugal: the 1998 International Exhibition (known as Expo 1998) was implemented on a large disused land in the eastern part of Lisbon. This area, now known as Parque das Nações (Fig. 5.3), was redeveloped after Expo 1998 and has become an urban area with commercial spaces, offices, homes and transport hubs (Gare do Oriente) integrated with the green spaces, which continues to attract many people.

---

devastation, desolate, devastated, plundered. It therefore refers to the characteristics of a space, but it is also the root of verbs such as devastating, destroying, deserting, external actions that act on spaces, bringing them to the condition of vastus. In the English language, the Latin word vastus, becomes vast, where it mainly expresses the meaning of greatness, but it can also be declined in the term vain (fourteenth-century lemma, from the ancient French vein) with the meaning of emptied of real, useless value, without profit, and in the term wane in the sense of left abandoned. The word waste shares the etymological origin with the word lemma, but acquires a wider meaning in the English language, making it more explicit what the induced characteristics of the object are and taking on the meaning of desert, desolate, uninhabited, uncultivated, unproductive.

Fig. 5.1 The Docks Libres district, Marseille

Fig. 5.2 View of the demolition of the degraded area of Toulon

Another example is that of disused airfields, which provided the opportunity to develop visions and test ideas for the design and use of future urban parks. In this process, abandoned airports have also become design and planning workshops where reinterpreting ancient dichotomies between public and private, practical and voluptuous purposes specific to the history of the Western landscape (Dümpelmann 2020). In recent decades, one of the most significant conversions in Europe has

Fig. 5.3 Parque das Nações, Lisbon

Fig. 5.4 A pasture near Tempelhof airport. Picture of Gert Schütz, April 1951 (in F. Panzini, Prati urbani. I prati collettivi nel paesaggio della città, cit., p. 218)

been the transformation of the former Berlin-Tempelhof airport into a new urban public park (Fig. 5.4). This metamorphosis occurred with new participatory planning methods and design processes such as open-source urbanism, the drafting of a flexible masterplan and informal revitalization, practices that intend to overcome the two visions considered sometimes irreconcilable of nature and public spaces (Mitchell 1995).

Viganò (1999) highlighted to us the roots of a progressive awareness of architects and urban planners about the growing inversion of the relationship between full and empty linked to the processes of modification of the contemporary city. From Colin

Rowe to Frank Lloyd Wright of "Broadacre city", up to the projections of the city-region and the results of the urban explosion in which the void becomes a structuring element of the city and the territory. The need to deal with this reversal today is not only attributable to an update of the traditional morphological categories of the soil project, nor only to a need to "live in nature", in which open spaces become the "natural" complement of new settlement forms. Rather, it is a radical reversal of attention and priority compared to the traditional construction of the city where buildings and fabric rules have always been privileged, while open spaces have reduced to being, at best, residual parts also from meaning and meaning (Secchi and Vigano 2011). Now it is precisely the emptiness and its high environmental, economic and social potential that play the main role in the landscape and value rethinking of cities.

The residues of the metropolises no longer appear as simple forms of a disengagement that testifies to a loss of power of the "manager" on his "territory", but as a component of the urban plans, as a support for agricultural activity, a new dimension now accepted from the revised notion of "land occupation" (Clément and Pieri 2014). The macro and microporosities of the urban sprawl are crossed by flows and expectations of different consistency and origin, cluttered by specialized or uncertain and transient uses and the wrecks of rapidly disappeared economies, by fragile ecological systems produced by the decisional fragmentation of urban mechanisms and by the use irrational resources.

In the void, density[3] is the protagonist of landscape in its most dynamic meaning, a promising fertile field for the convergence of a multiplicity of questions and actions that work on the relationships between flows and places. It is where they interact, juxtapose and often conflict the mosaic of rural–urban spaces and periurban the pattern of surface and deep waters, the system of infrastructure networks and the spread of waste and waste areas. These landscape networks contribute to rethinking the shape, the ecology and the "public" offer of spaces in the city, giving new life to the intuition of the "soil project" of the 1980s. The networks of blue, green and slow infrastructures are also able to produce a progressive metamorphosis of the different settlement patterns into landscape matrices, hosting the widespread devices of their ecological regeneration and actively participating in the recycling of scarce resources. This configures dross capes no longer as "black holes" but as essential and priority materials for the activation of tactics anchored in this strategic perspective (Salvati and Ricciardo Lamonica 2020).

The recycling strategy, therefore, focuses on the construction of "new soils" capable of creating over time a connective system of multifunctional open spaces,

[3] One area to investigate in dealing with the theme of density is certainly that of proxemics. The term, coined in the 1960s by the American anthropologist Edward T. Hall, derives from the Latin proximus (neighbor) and the Greek séma (sign), and indicates a discipline that studies what personal and social space are and how man perceives them. In semiology it is a sector that deals with the use and management that man makes of space, when he distances between individuals, in order to bring them closer or further away in daily interactions and in the structuring of living and urban spaces. The architecture of these spaces and the density of use of them takes into account the proxemic distances of the elements of the urban landscape (Reale 2008).

landscape networks for the city and the territory, strongly characterized by the regeneration of devices and water, energy, waste and mobility networks. A qualitative strategic-adaptive design approach that goes beyond the sectoral practices traditionally used (for example remediation), identifying itself as a constitutive element of an urban project, as well as an ecologically oriented landscape (Pili et al. 2017).

The need to carry out substantial replacement interventions, in consideration of the high level of building "consumption" and the consequent biophysical degradation of the urban and peri-urban territory, represents an area still to be explored in the design, as well as the need to frame recovery interventions in a wider system of plan coherences. We have seen how in the metropolises the areas of the so-called drosscapes often offer a punctual but dense repertoire of opportunities for transformation. Another important issue is that of research in the design of an effective functional mix. This provides a plurality of users, uses, relationships capable of restoring conditions of welcome, vitality and safety typical of a liveable city, but also offering greater flexibility and durability of transformative operations, avoiding the harmful consequences produced by the crisis and the sudden abandonment of a single type of economic activity and a single functional destination. Another consideration regards the qualitative role that the soil project can attribute to the most resistant and symbolic part of a metropolitan area, the open spaces in their various forms, offering greater perspectives of stability and renewal of the meaning and meaning of urban places, faced with the questions of functional change (De Rosa and Salvati 2016).

The new territorial structure of the metropolis, containing more or less ample continuity solutions, has open spaces that can be used in part in agricultural production and allow a better organization and networking of parks, woods and equipped areas. European metropolises are transforming themselves, organizing themselves on larger reference scales according to a logic of polycentrisms and networks in which agricultural spaces can become as precious as monumental gardens or historic centers already are (Mininni and Donadieu 2013).

The landscape planning season explored in the available plans and on different levels of scales and regulatory tools the possibilities of urban policies to develop a common program with agricultural ones, involving agricultural spaces in the urban planning project. The urban campaign opened the urban project to a comparison with a material, such as the agricultural space, which brings with it its symbols, values and ecologies. In fact, one of the main transformations that have taken place in recent decades, concerning European metropolises, is the city-countryside relationship and with it a new landscape. These are signs of new ecologies that are emerging between territory and society, partly dependent on urban culture and from the rural one, but in many respects bearers of an unprecedented proposal of sustainability and new forms of spatiality on which it is worth questioning (Colantoni et al. 2015).

For Mininni and Donadieu (2013), it is the city that should take care of protecting the countryside, ensuring the permanence of the void, the bearer of values of nature, proposing agricultural activity. The challenge of the contemporary city could start precisely from these peri-urban agricultural spaces, whose urbanity is strengthened the more the urban centers bring their margins closer, incorporating countryside areas. In these places, urban services and new peripheral centralities are increasingly concentrated, but more creative forms of agriculture are developing or would have the potential to emerge, innovative activities practiced by both citizens and farmers, who increasingly receive the advantages that proximity to the city, can offer.

Given the urban concentration expected in 2050 (the year in which 70% of the world's population will live in cities), it could perhaps be agreed that it would be more realistic to build by including agricultural spaces and farmers in the urban region, rather than just looking for the densification of the city and neighborhoods, and rejecting agriculture toward the suburbs (Cecchini et al. 2019). The latter could indeed be considered by planners as an urbanization tool capable of lastingly organizing the territory of cities, renewing itself to meet the needs of citizens, both from a production and recreational point of view. In reality, as Donadieu explains well in his book "Urban campaigns. A new proposal for the city landscape", it would be a matter of reconstructing historically existing relationships between cities and agriculture.

> For this new campaign to become a landscape, and consequently acquire a relative perennially, it is evident that it must be created with those visible qualities it does not have. It is the function of a landscape project that of designating and building the landscape structures that will constitute the eco-symbols: hedges, waterways, bridges, small woods, villages, orchards, chapels, borders, etc.[4]

The potential of the sedimented and millennial work of the earth combines economy and soil care with the survival of the landscape. It is a theme that is increasingly used in territorial studies to interpret and regenerate the settled area, its voids, in the form of a rural and agricultural landscape (Perrin et al. 2018). Agriculture also makes it possible to reinterpret the settlement and housing problems inherent in the phenomenon of dispersion, through the search for values and plots sedimented in the urban countryside, proposing new forms of habitability for a fragmented territory such as the "periurban" (Donadieu 2006). Nature-urban hybridization today therefore represents a key issue. The quality of the habitats could be achieved by using the characteristic agricultural and forest environments to graft the new urban structures (Fig. 5.5).

The urban voids and green spaces, thanks to the urban landscape project, could become the new plots and new patterns on which to base the organization of urban countryside. It considers and manages them no longer as inert spaces that would isolate the nodes and central urban areas, but as territories, agricultural and wooded, living, in slow or rapid, cyclical or continuous becoming. In a world where cities are getting bigger and more congested, public spaces are gaining interest. Moreover, social transformations have brought about a change in the use of urban green (Di Feliciantonio et al. 2018). The intrinsic value of these spaces has been recognized, and should always be, in the sense of urban environmental protection (Szlavecz et al. 2011). It would also be desirable to preserve as much as possible the intrinsic dynamic and procedural aspect of the vegetation, using the spontaneous one (Fig. 5.6), to:

[4] P. Donadieu, P. De Stefano, M. Mininni, *Urban Campaigns: A New Proposal for the City Landscape,* Donzelli 2006, p. 85.

Fig. 5.5 Parc de la Lironde, Montpellier. Transposition of a part of the typical landscape of wetlands within a new neighborhood, through the creation of two basins for collecting rainwater, transformed into a pond. This 4.5-hectare park was inaugurated in 1995 and is entirely crossed by the Lironde stream

(i) protect and encourage soil biodiversity;
(ii) favor climate adaptation;
(iii) improve the aesthetic quality of urban areas and enhance their attractiveness by rejecting dispersion.

It is essential to work for a city where nature and urbanity converge and reinforce each other, where green infrastructures (Fig. 5.7) reach connectivity and where green heritage reaches continuity with the natural area that surrounds it. The first planning experiences developed right within an ideology aimed at re-establishing an organic relationship between the environment and urbanization processes.

For example, the City of Barcelona and its metropolitan area implement specific policies to allow nature to adapt to the city and improve biological diversity, also considering that in a metropolis with greater green infrastructure, people can benefit from higher levels of health and well-being (Figs. 5.8 and 5.9).

Fig. 5.6 Urban wildness: grove in Park am Gleisdreieck, Berlin 2011. Photo by Norbert Kühn (in F. Panzini, Prati urbani. I prati collettivi nel paesaggio della città, cit., p. 145)

Fig. 5.7 Frederick Law Olmsted's project for an "emerald necklace" (Emerald Necklace, Boston, 1878–1896), that is, a network of public green spaces, is considered by some urban planners to be one of the first examples of green infrastructure (The term green infrastructure appeared in the United States in the mid-90s, but Benedict and McMahon indicate it as the first inventor FL Olmsted, who already around 1900 hoped for the connection of urban green areas to improve the quality of life in cities. With Olmsted the debate on parks and the idea of building a new civilization in empathy with nature has been transformed from a literary utopia into a real political and institutional problem, into new perspectives for urban social reform) (in F. Panzini, Prati urbani. I Prati collettivi nel paesaggio Della città, cit., p. 82)

Fig. 5.8 The network of urban green corridors in Barcelona connects the green spaces within the city to the four main natural areas that surround it: the Collserola mountain range, the coast, the Besòs River and the Llobregat River

Fig. 5.9 In 2013, the city of Barcelona launched the 2020 Green Infrastructure and Biodiversity Plan, pledging to "preserve and enhance the natural heritage of the city" and allow "everyone [...] to benefit from it". Image is taken from website https://www.iucn.org/content/barcelonas-green-infrastructure-and-biodiversity-plan

Habitability and, therefore, the improvement of the quality of the peri-urban territories has become a goal that cannot be longer postponed, both to accommodate agricultural businesses and inhabitants. Living in the agricultural and forestry space presupposes both a moral and aesthetic project that transforms it, as Donadieu himself affirmed into "an urban countryside".

References

Antrop, M (2004) Landscape change and the urbanization process in Europe. *Landscape and Urban Planning* 67. Development of European Landscapes:9–26. https://doi.org/10.1016/S0169-2046(03)00026-4.

Bajocco S, De Angelis A, Salvati L (2012) A satellite-based green index as a proxy for vegetation cover quality in a Mediterranean region. Ecol Ind 23:578–587. https://doi.org/10.1016/j.ecolind.2012.05.013

Bajocco S, Dragoz E, Gitas I, Smiraglia D, Salvati L, Ricotta C (2015) Mapping forest fuels through vegetation phenology: The role of coarse-resolution satellite time-series. PLOS ONE 10: e0119811. https://doi.org/10.1371/journal.pone.0119811

Bajocco S, Salvati L, Ricotta C (2011) Land degradation versus fire: A spiral process? Progress in Physical Geography: Earth and Environment 35:3–18. https://doi.org/10.1177/0309133310380768

Barbati A, Corona P, Salvati L, Gasparella L (2013) Natural forest expansion into suburban countryside: Gained ground for a green infrastructure? Urban Forestry & Urban Greening 12:36–43. https://doi.org/10.1016/j.ufug.2012.11.002

Berger, A (2006) Drosscape: Wasting Land in Urban America. Princeton Architectural Press

Biasi R, Colantoni A, Ferrara C, Ranalli F, Salvati L (2015) In-between sprawl and fires: long-term forest expansion and settlement dynamics at the wildland–urban interface in Rome, Italy. Int J Sust Dev World 22:467–475. https://doi.org/10.1080/13504509.2015.1064488

Cecchini M, Zambon I, Pontrandolfi A, Turco R, Colantoni A, Mavrakis A, Salvati L (2019) Urban sprawl and the 'olive' landscape: sustainable land management for 'crisis' cities. GeoJournal 84:237–255. https://doi.org/10.1007/s10708-018-9848-5

Chas- ML, Touza J, García-Martínez E (2013) Forest fires in the wildland–urban interface: A spatial analysis of forest fragmentation and human impacts. Appl Geogr 43:127–137. https://doi.org/10.1016/j.apgeog.2013.06.010

Clément, G, De Pieri, F (2014) Manifesto del Terzo paesaggio. Translated by G. Lucchesini. Macerata, Italy: Quodlibet

Clemente M, Zambon I, Konaxis I, Salvati L (2018) Urban growth, economic structures and demographic dynamics: exploring the spatial mismatch between planned and actual land-use in a Mediterranean city. Int Plan Stud 23:376–390. https://doi.org/10.1080/13563475.2018.1500277

Colantoni A, Mavrakis A, Sorgi T, Salvati L (2015) Towards a 'polycentric' landscape? Reconnecting fragments into an integrated network of coastal forests in Rome. Rendiconti Lincei 26:615–624. https://doi.org/10.1007/s12210-015-0394-5

Corona P, Chirici G, Marchetti M (2002) Forest ecosystem inventory and monitoring as a framework for terrestrial natural renewable resource survey programmes. Plant Biosystems - an International Journal Dealing with All Aspects of Plant Biology 136:69–82. https://doi.org/10.1080/11263500212331358531

Cuadrado-Ciuraneta S, Durà-Guimerà A, Salvati L (2017) Not only tourism: unravelling suburbanization, second-home expansion and "rural" sprawl in Catalonia, Spain. Urban Geogr 38:66–89. https://doi.org/10.1080/02723638.2015.1113806

De Rosa S, Luca S (2016) Beyond a 'side street story'? Naples from spontaneous centrality to entropic polycentricism, towards a 'crisis city.' *Cities* 51. Current Research on Cities:74–83. https://doi.org/10.1016/j.cities.2015.11.025

Di, Feliciantonio, Cesare LS, Sarantakou E, Rontos K (2018) Class diversification, economic growth and urban sprawl: evidences from a pre-crisis European city. Qual Quan 52:1501–1522. https://doi.org/10.1007/s11135-017-0532-5

Di Simine D, Pileri P, Ronchi S (2013) Consumo di suolo: una questione ambientale. FORUM Di Lexambiente 13:14–23

Dümpelmann, S (2020) Prati urbani/City meadows. Treviso, Italy: Antiga Edizioni. http://www.fbsr.it/edizioni/prati-urbani-city-meadows/. Accessed September 30

Duvernoy I, Zambon I, Sateriano A, Salvati L (2018) Pictures from the other side of the fringe: Urban growth and peri-urban agriculture in a post-industrial city (Toulouse, France). J Rural Stud 57:25–35. https://doi.org/10.1016/j.jrurstud.2017.10.007

Elia M, Lafortezza R, Colangelo G, Sanesi G (2014) A streamlined approach for the spatial allocation of fuel removals in wildland–urban interfaces. Landscape Ecol 29:1771–1784. https://doi.org/10.1007/s10980-014-0070-7

Glæsner N, Helming K, De Vries W (2014) Do current European policies prevent soil threats and support soil functions? Sustainability 6:9538–9563. https://doi.org/10.3390/su6129538

Healey P (2004) The treatment of space and place in the new strategic spatial planning in Europe. Int J Urban Reg Res 28:45–67. https://doi.org/10.1111/j.0309-1317.2004.00502.x

Jurgens, Clifford R (1993) Strategic planning for sustainable rural development. *Landscape and Urban Planning* 27. Special Issue New Strategies for Sustainable Rural Development:253–258. https://doi.org/10.1016/0169-2046(93)90058-L

Lamonica GR, Recchioni MC, Chelli FM, Salvati L (2020) The efficiency of the cross-entropy method when estimating the technical coefficients of input–output tables. Spat Econ Anal 15(1):62–91

Li X, Pan J, ed. (2014) China green development index report 2012: Regional comparison. Current Chinese Economic Report Series. Berlin Heidelberg: Springer-Verlag. https://doi.org/10.1007/978-3-642-54178-0

Mininni M, Pierre D (2013) Campagne urbane. Una nuova proposta di paesaggio della città. Translated by M. L'Erario. Roma: Donzelli

Mitchell D (1995) The end of public space? People's park, definitions of the public, and democracy. Ann Assoc Am Geogr 85:108–133. https://doi.org/10.1111/j.1467-8306.1995.tb01797.x

Munafò M, Salvati L, Zitti M (2013) Estimating soil sealing rate at national level—Italy as a case study. Ecol Ind 26:137–140. https://doi.org/10.1016/j.ecolind.2012.11.001

Oliveira E, Hersperger AM (2018) Governance arrangements, funding mechanisms and power configurations in current practices of strategic spatial plan implementation. Land Use Policy 76:623–633. https://doi.org/10.1016/j.landusepol.2018.02.042

Oliveira E, Tobias S, Hersperger AM (2018) Can strategic spatial planning contribute to land degradation reduction in urban regions? State of the art and future research. Sustainability 10:949. https://doi.org/10.3390/su10040949

Perrin C, Nougarèdes B, Sini L, Branduini P, Salvati L (2018) Governance changes in peri-urban farmland protection following decentralisation: A comparison between Montpellier (France) and Rome (Italy). Land Use Policy 70:535–546. https://doi.org/10.1016/j.landusepol.2017.09.027

Pili S, Grigoriadis E, Carlucci M, Clemente M, Salvati L (2017) Towards sustainable growth? A multi-criteria assessment of (changing) urban forms. Ecol Ind 76:71–80. https://doi.org/10.1016/j.ecolind.2017.01.008

Prokop P, Kruczkowska B, Syiemlieh HJ, Bucała-Hrabia A (2018) Impact of topography and sedentary swidden cultivation on soils in the hilly uplands of North-East India. Land Degrad Dev 29:2760–2770. https://doi.org/10.1002/ldr.3018

Reale, L (2008) Densità, città, residenza. Tecniche di densificazione e strategie anti-sprawl. Roma: Gangemi

Recanatesi F, Clemente M, Grigoriadis E, Ranalli F, Zitti M, Salvati L (2016) A fifty-year sustainability assessment of italian agro-forest districts. Sustainability 8:32. https://doi.org/10.3390/su8010032

Salvati L (2013) Urban containment in action? Long-term dynamics of self-contained urban growth in compact and dispersed regions of southern Europe. Land Use Policy 35:213–225. https://doi.org/10.1016/j.landusepol.2013.05.009

Salvati L (2014a) Towards a polycentric region? The socio-economic trajectory of Rome, an 'Eternally Mediterranean' city. Tijdschr Econ Soc Geogr 105:268–284. https://doi.org/10.1111/tesg.12054

Salvati L (2014b) Agro-forest landscape and the 'fringe' city: A multivariate assessment of land-use changes in a sprawling region and implications for planning. Sci Total Environ 490:715–723. https://doi.org/10.1016/j.scitotenv.2014.05.080

Salvati L, Carlucci M (2016) The way towards land consumption: Soil sealing and polycentric development in Barcelona. Urban Studies 53:418–440. https://doi.org/10.1177/0042098014563486

Salvati L, Ferrara A, Chelli F (2018) Long-term growth and metropolitan spatial structures: an analysis of factors influencing urban patch size under different economic cycles. Geografisk Tidsskrift-Danish Journal of Geography 118:56–71. https://doi.org/10.1080/00167223.2017.1386582

Salvati L, Lamonica, GR (2020) Containing urban expansion: Densification vs greenfield development, socio-demographic transformations and the economic crisis in a Southern European City, 2006–2015. Ecol Ind 110: 105923. https://doi.org/10.1016/j.ecolind.2019.105923

Salvati L, Zambon I, Chelli FM, Serra P (2018) Do spatial patterns of urbanization and land consumption reflect different socioeconomic contexts in Europe? The Science of the Total Environment 625:722–730. https://doi.org/10.1016/j.scitotenv.2017.12.341

Satterthwaite D, McGranahan G, Tacoli C (2010) Urbanization and its implications for food and farming. Philosophical Transactions of the Royal Society B: Biological Sciences 365:2809–2820. https://doi.org/10.1098/rstb.2010.0136

Scott M (2006) Strategic spatial planning and contested ruralities: Insights from the republic of Ireland. Eur Plan Stud 14:811–829. https://doi.org/10.1080/09654310500496149

Secchi, and Vigano (2011). La Ville Poreuse: Un Projet Pour le Grand Paris et La... Genève: MētisPresses

Szlavecz, Katalin, Paige Warren, and Steward Pickett (2011) Biodiversity on the urban landscape. In: Richard P, Cincotta and Larry J Gorenflo (eds) Human population: Its Influences on biological diversity. Ecological Studies. Berlin, Heidelberg: Springer, pp 75–101. https://doi.org/10.1007/978-3-642-16707-2_6

Tobias S (2013) Preserving ecosystem services in urban regions: Challenges for planning and best practice examples from Switzerland. Integr Environ Assess Manag 9:243–251. https://doi.org/10.1002/ieam.1392

Tomao A, Quatrini V, Corona P, Ferrara A, Lafortezza R, Salvati L (2017) Resilient landscapes in Mediterranean urban areas: Understanding factors influencing forest trends. Environ Res 156:1–9. https://doi.org/10.1016/j.envres.2017.03.006

Viganò, Paola (1999) La città elementare. Skira

Chapter 6
Conclusions

Abstract Soil protection policies are evolving in Europe, with the Member States at different stages of development in developing possible strategies. As suggested in the present book, a key concept that could be more integrated into policies aimed at achieving more sustainable urban forms is that of quality: a concept that applies to the soil (concerning its intrinsic, therefore biophysical, properties, and its dynamic properties, that is, proper to the interaction of the different uses of the soil with the landscape), but which in the end goes to incorporate ecosystems and their state of health. This means that land quality is not only the expression of an indicator that measures certain environmental parameters, but it is also a concept that could allow us to consider together the management and cultural aspects of actions aimed at improving landscape integration and which are at the simultaneously aimed at the recovery and enhancement of soil functions. Land quality has allowed in this research to reason on how the urbanization pattern (compact, dispersive and intermediate) affects land degradation in 76 major European cities in a diversified way. They cover an area—the Mediterranean basin—which from an ecological point of view is very important and interesting: it is a biodiversity hotspot and is particularly sensitive to land degradation. The understanding of land degradation processes and their interaction with urbanization is certainly a good starting point for the improvement of land quality in the metropolitan areas considered, but, as well stressed several times in this study, it must necessarily be accompanied by a change of vision that considers the soil as a living entity, expanding the interpretative categories to be used in planning processes. It was also stressed that spatial planning should also be more practice-oriented in terms of actions against land degradation, for the creation of truly sustainable systems, using specific data on land use and quality and opening up more to the project of the landscape.

Keywords Land quality · Land degradation · Urban sprawl · Mediterranean metropolis

I. Tombolini et al., *Land Quality and Sustainable Urban Forms*, Springer Geography,
https://doi.org/10.1007/978-3-030-94732-3_6

6.1 Final Remarks

The new socio-spatial structure of the city, moving away from the relationship of strength between the center and the periphery, reinforces a profile of fragmentation, isolation and fractality that comes forward in contemporary metropolitan areas, where it is increasingly important to design and build sustainable systems, considering, explicitly, urbanization as part of the solution to achieving regional and global sustainability. Former UN Secretary-General Kofi Annan recently passed away, expressed this concept well at the Conference on sustainable urban development held in Moscow on 5 June 2002: "The future of humanity lies in the cities".

The most evident change that is taking place in the landscape of European metropolises is the explosion of the city: this phenomenon not only has very important social implications, allowing to extend economic, cultural and friendly urban relationships beyond the conventional limits of the city but also impacts on a structural level and with the environmental sphere.

Given the cross-sectoral nature of soil-related issues and given the diversity of environmental and socio-economic pressures to be taken into account also at the level of governance across Europe, it is not surprising that there are many different political instruments for soil protection, both at Union level European that at the level of the individual Member States. The maintenance of the ecological functionality of the soil in most cases is only implicitly considered in the legislation of the Member States. The lack of a common and integrated strategic framework represents an important shortcoming, which could be filled by the proposal for a framework directive on soil, which was later withdrawn. Therefore there is still no political strategy which, in an integrated way, can:

- clarify and unequivocally define issues relating to soil (which concern for example its state of health);
- establish priorities for action and objectives;
- define the parameters for monitoring;
- define the role of the various political instruments in ensuring the ecological functionality of the soil.

In the absence of a common political framework, it has not yet been possible to define an address through which to establish agreed objectives. This not only means that the existing policy framework has limitations for soil protection, but also that the strengths and opportunities identified cannot be fully explored and exploited. The importance of achieving this integration is underlined by the fact that the European Union's national and regional policies interact with international ones, such as the United Nations Convention against desertification. The emphasis placed on stopping and reversing land degradation in the Sustainable Development Goals, and achieving land degradation neutrality (Sustainable Development Goal 15.3) should also offer opportunities to emphasize soil protection in Europe.

In addition to the lack of strategic coordination, the investigated policies present two other important weaknesses:

- soil protection is often an implicit result as it mainly derives from the protection of other environmental resources, from the consideration of other environmental criticalities or the achievement of other objectives;
- the policies that offer a strategic vision are not binding and, as such, cannot be used as key elements for the integration and strengthening of soil protection within existing EU laws.

Soil protection policies are evolving in Europe, with the Member States at different stages of development in developing possible strategies. As suggested in the present book, a key concept that could be more integrated into policies aimed at achieving more sustainable urban forms is that of quality: a concept that applies to the soil (concerning its intrinsic, therefore biophysical, properties, and its dynamic properties, that is, proper to the interaction of the different uses of the soil with the landscape), but which in the end goes to incorporate ecosystems and their state of health.

This means that land quality is not only the expression of an indicator that measures certain environmental parameters, but it is also a concept that could allow us to consider together the management and cultural aspects of actions aimed at improving landscape integration and which are at the simultaneously aimed at the recovery and enhancement of soil functions.

Land quality has allowed in this research to reason on how the urbanization pattern (compact, dispersive and intermediate) affects land degradation in 76 major European cities in a diversified way. They cover an area—the Mediterranean basin—which from an ecological point of view is very important and interesting: it is a biodiversity hotspot and is particularly sensitive to land degradation.

The results showed that dense and semi-dense settlements represent more sustainable urban forms than discontinuous settlements, more preserving the land quality of the surrounding agricultural areas and, consequently, the environmental services they provide and the biodiversity associated with these spaces.

Research certainly represents an expansion of knowledge: among the three forms of urbanization considered, the discontinuous one has consumed more frequently soils with better land quality. Perhaps the evidence is provided by noting that the built areas progressively occupy the most productive and fertile soils, but overall quantitative and qualitative estimates have also been provided, including numerous metropolitan areas, which were not yet known. The research could also have an application aspect: the methodological framework developed, based on freely and easily accessible data, continuously updated, could be integrated into an interoperable geographic information system to contribute to the orientations of future planning.

From the union of the application aspect and the knowledge contribution, further reflections could be developed at the level of planning and also of the landscape project, enhancing the sharing of good practices, the dissemination of tools and technologies to local actors and increasingly refining the techniques of monitoring through understanding the continuous evolution of land degradation processes, and if possible, the construction of even more effective indicators thanks to the continuous improvements that satellite data offer from information detail. The understanding of land degradation processes and their interaction with urbanization is certainly a good starting point for the improvement of land quality in the metropolitan areas considered, but, as well stressed several times in this study, it must necessarily be accompanied by a change of vision that considers the soil as a living entity, expanding the interpretative categories to be used in planning processes. It was also stressed that spatial planning should also be more practice-oriented in terms of actions against land degradation, for the creation of truly sustainable systems, using specific data on land use and quality and opening up more to the project of the landscape.

Glossary

Land Ecologically multifunctional system, whose natural capital, consisting of the soil and the biodiversity associated with it, interacting with water and the atmosphere, generates a flow of ecosystem services that support human well-being, ensuring the life and sustenance of individuals and communities, animals and plants (UNCCD, paragraph 22). This system includes the soil but also consists of many other interacting dimensions.

Soil The upper layer of the earth's crust, consisting of mineral particles, organic matter, water, air and living organisms, which represents the interface between earth, air and water and houses a large part of the biosphere. Soil is an essentially non-renewable resource and a very dynamic system, which performs numerous functions and provides essential services for human activities and the survival of ecosystems (from the Thematic Strategy of the European Union for the protection of soil elaborated by the European Commission in 2006).

Ecosystem services Multiple benefits provided by ecosystems to mankind (definition given by the Millennium Ecosystem Assessment 2005). The Millenium Ecosystem Assessment describes four categories of ecosystem services: life support, regulation, supply, cultural values.

Ecosystem services	Soil-functions
Support services (necessary for the production of all other ecosystem services)	
Soil formation	Weathering of primary minerals and release of nutrients; modification of soil texture Transformation and accumulation of organic matter Creation of structures (aggregates, horizons) for the flow of gas and water and the growth of the roots Formation of surfaces considers the retention and ion exchange
Primary production	Soil for seed germination and root growth Storage and supply of nutrients and water for plants

(continued)

I. Tombolini et al., *Land Quality and Sustainable Urban Forms*, Springer Geography,
https://doi.org/10.1007/978-3-030-94732-3

(continued)

Ecosystem services	Soil-functions
Nutrient cycle	Transformation and mineralization of organic matter by soil organisms Conservation and release of nutrients from surfaces
Regulatory services: benefits obtained from the regulation of ecosystem processes	
Regulation of water quality	Filtration of substances in soil water
Regulation of water supply	Regulation of infiltration and flow of water in the soil Drainage of water
Climate regulation	Regulation of CO_2, N_2O and CH_4 emissions Carbon sequestration
Erosion regulation	Soil conservation on the earth's surface
Supply services: products ("goods") of ecosystems from which man directly benefits	
Food supply	Supply of water, nutrients, physical support for plants consumed by humans and animals
Water supply	Water retention and purification
Supply of fibers and raw materials	Supply of water, nutrients and physical support for the development of plants for bioenergy and fiber
Shelter	Habitat supply for soil-dwelling animals, birds, etc.
Genetic resources	Source of biological materials potentially or really important from a genetic point of view
Cultural services: material benefits through spiritual enrichment, aesthetic- perceptual experiences, heritage conservation and recreational activities	
Aesthetic and spiritual functions	Conservation of the diversity of the natural and cultural landscape

Source Adapted from FAO e ITPS (2015)

Resources Properties and Status

Land quality Set of intrinsic qualities (permanent characteristics difficult to modify such as depth) and dynamic qualities of the soil (provision of ecosystem services) which guarantee the productivity of natural and semi-natural systems. It can be considered, as is done in this thesis, at the antipodes of land degradation.

Resilience The concept originated in the field of ecology to indicate the multitude of changes brought about by the disturbance and the post-disturbance restoration dynamics. Holling (1973) popularized this term in the more general context of "ecosystem stability" by defining "resilience" as the amount of disturbance that an ecosystem can support without changing the fundamental processes of self-organization, its structure and functions. The concept of "ecological resilience" is generally defined at the ecosystem level but sociologists have also introduced the concept of "social resilience", usually formulated in terms of the ability of social groups or communities to respond to externally induced shocks (such as extreme weather events, epidemics, large infrastructure projects). This led to the formulation of the concept of "socio-ecological resilience" (Adger et al. 2005) to link anthropic action, social institutions and market dynamics to strategies for the management of natural resources. The extension to socio-ecological systems allows to deal with issues related to the formation of new paths, the ability to reorganize and therefore to renew a system (Folke et al. 2010).

Sensitivity Degradation level reached in a certain territory due to the scalar processes that generate desertification as the last step, such as climate change, soil erosion, deforestation and salinization, triggered by natural or anthropic causes (from McCarthy's IPCC report et al. 2001). The effect can be direct (a change in agricultural productivity in response to a change in the average, in the range or the variability of temperature) or indirect (damage caused by an increase in the frequency of flooding).

Vulnerability Level of susceptibility (or resilience) of a system to the phenomena of desertification (from the IPCC report by McCarthy et al. 2001). It depends on the sensitivity of the system and the ability to adapt to the variation of certain conditions, for example, climatic conditions.

Processes

Soil consumption A phenomenon associated with the loss of a fundamental environmental resource, due to the occupation of originally agricultural, natural or semi-natural surface, and is defined as a variation from a non-artificial cover (unconsumed soil) to an artificial cover of the soil (consumed soil) (ISPRA SNPA 2018). Land use (i.e. the new artificial cover) can be divided into two main categories.

- Permanent soil consumption: buildings, buildings; asphalted roads; railway headquarters; airports (runways and waterproof/paved handling areas); ports (docks and waterproof/paved handling areas); other waterproof/paved areas; permanent paved greenhouses; landfills.
- Reversible soil consumption: dirt roads; construction sites and other areas in beaten earth (squares, parking lots, courtyards, sports fields, permanent material storage); non-renaturalized extraction areas; quarries in the pitch; photovoltaic fields on the ground; other artificial coverings whose removal restores the initial soil conditions.

Desertification Serious and irreversible form of land degradation (UNCCD 1994), which occurs in arid, semi-arid and dry sub-humid areas due to many factors, including climatic variations (prolonged drought periods) and human activities (such as the removal or over-exploitation of the upper arable soil layer).

Landscape fragmantation A dynamic process, of anthropic origin, which consists in the division of the natural environment into increasingly smaller and isolated fragments or patches—unconsumed areas without significant artificial elements that fragment them interrupting their continuity—separated by a soil cover matrix transformed by the man. The fragmentation effect produced in an environmental mosaic varies about the characteristics of the matrix. The destruction and transformation of natural environments, their reduction and increased insulation are the main components of the fragmentation process. They influence the structure and dynamics of certain sensitive animal and plant populations, to the point of altering community parameters, ecosystem functions and ecological processes. This process is mainly the result of urban expansion phenomena that take place according to more or less sustainable forms and of the development of the infrastructure network aimed at improving the connection of urbanized areas through linear works, thus generating effects of the reduction of the continuity of ecosystems, habitats and landscape units (ISPRA SNPA 2018).

Land degradation Decrease or disappearance of biological or economic productivity and the complexity of non-irrigated cultivated land, irrigated cultivated land, pastures, forests or wooded areas following the use of the lands or due to one or more phenomena, due to the activity of man and his ways of settling (Dodds et al. 2016).

At the heart of this definition are the functions and the economic value of the earth for agriculture and forestry, which are promoted by human activities and manifest themselves negatively in the form of measurable phenomena such as erosion, loss of soil quality and the loss of vegetation. Other institutions such as the Intergovernmental Science-Policy Platform on Biodiversity and Ecosystem Services and the Global Environment Facility provide a broader and less specific interpretation of soil degradation: they consider it more a loss or a reduction of ecosystem services and functions and do not differentiate between human activities or natural degradation processes.

Soil sealing Permanent coverage of part of the land and its soil with artificial materials (such as asphalt or concrete) for the construction, for example, of buildings and roads or other structures made of partially or completely waterproof material. In general, a part of the settlement area is waterproofed, except gardens, urban parks and other green spaces.

Urban sprawl Spread of low density dispersed settlements from the urban center to the outside, which derives essentially from an unplanned urban development and determines a mix of different types of land use in the urban and peri-urban fringe of many important cities (European Agency for Environment 2006). This is a traditional phenomenon well studied for the cities of North America (Bruegmann 2006) in place since the beginning of the twentieth century, while it is relatively recent as regards European cities (Couch et al. 2007).

Programs and Products for the Interpretation of Properties and Processes

Copernicus Launched in 1998 by the European Commission and a pool of space agencies, the Copernicus European Earth Observation Program, formerly known as GMES (Global Monitoring for Environment and Security), is a complex set of systems that collects information from multiple sources, i.e. Earth observation satellites and land, sea and airborne sensors. It integrates and processes all this information, providing users, institutional and related to the industry sector, with reliable and updated information through a series of services that concern the environment, the territory and safety. Copernicus also has among its objectives to guarantee Europe substantial independence in the collection and management of data on the state of health of the planet, supporting the needs of European public policies through the provision of precise and reliable services. This information relates to six thematic areas: land, sea, atmosphere, climate change, emergency management and security.

The "Copernicus Land Monitoring Service" deals with environmental terrestrial applications, which has four main components: global, pan-European, local, data and reference images.

DISMED (Desertification Information System for the Mediterranean) Project formulated by the UNCCD Secretariat in collaboration with the European Environment Agency and the Italian Foundation of Applied Meteorology (FMA), which aims to provide an action framework for sustainable development in arid, semi-arid and dry sub-humid, and to combat desertification and the effects of drought. This mechanism is based on the exchange of information, the circulation of data and the definition of a common information system to monitor the physical and socio-economic conditions of areas sensitive to desertification.

MEDALUS (Mediterranean Desertification And Land Use) International research project launched in 1996 to study the effects of desertification in Mediterranean Europe. It is funded by the Commission of the European Communities, Directorate General XII for Science, Research and Development within the environmental program. The main objective is to undertake and consolidate research that can contribute to a deeper understanding of desertification processes, applying the knowledge acquired to develop modeling tools and possible scenarios of desertification.

Urban Atlas Product of the local component of the Copernicus Land Monitoring service, which focuses and provides detailed information on sensitive areas and predisposed to changes from an environmental point of view (hotspots). Urban Atlas, in particular, provides a series of maps on the use and coverage of European soil at a scale of 1: 10,000, with a focus on urban areas. The thematic classes are based on the Corine Land Cover nomenclature and are distinguished between "urban", with a minimum cartographic unit of 0.25 hectares, and "non-urban", with a minimum cartographic unit of 1 hectare.

Land Initiatives

LDN (Land Degradation Neutrality) State in which the quantity and quality of resources associated with the land system, necessary to support ecosystem functions and services and to improve food security, remain stable or increase within a certain time and space scales (IAEG-SDGs 2016). The LDN concept was first introduced in the "Rio + 20", which was later taken up by the United Nations General Assembly in 2015 and included in target 15.3 of the Sustainable Development Goals.

PAN (National Action Program) The UNCCD identifies in the National Action Programs (PAN) the tools through which to adopt and implement actions to combat desertification based on the guidelines of resolution 32/172 of 19 December 1977 of the United Nations assembly. They were developed through a participatory approach that involved the relevant government offices, scientific institutions and local communities. The NAPs are also linked and contain synergies with other global conventions on climate, biodiversity and water protection.

SDGs Sustainable Development Goals (sustainable development objectives) The SDGs are 17 Sustainable Development Goals which include 169 targets. They are a set of objectives conceived and elaborated by the United Nations Organization for International Development; expired at the end of 2015, they have been revalidated for the period 2015–2030. For the land system and the soil, the objective 15 and the target 15.3 are of particular importance.

Goal 15 "Manage forests sustainably, combat desertification, halt and reverse land degradation and halt the loss of biodiversity".

Target 15.3 "By 2030, combat desertification, restore degraded lands and soil, including those affected by desertification, drought and floods, and commit to achieving a land degradation-neutral world" (UNDESA 2016).

UNCCD (United Nations Convention to Combat Desertification in those countries living serious darkness and/or desertification, in particular in Africa) Convention to combat desertification and mitigate the effects of drought through national programs incorporating long-term strategies supported by international cooperation and partnership agreements. The Convention, the only one that derives from a direct recommendation of Agenda 21 of the Rio Conference, was adopted in Paris, France, on 17 June 1994, and entered into force in December 1996. It is the only internationally binding framework set up to address the problem of desertification. The Convention is based on the principles of participation, partnership and decentralization.

Space Units

City The new definition of the city according to Eurostat (2015) is based on the presence of an urban center (spatial concept based on cells with high population density) and is divided into four phases:

- phase 1: all cells with a population density greater than or equal to 1,500 inhabitants per square kilometer are selected;
- phase 2: the high-density contiguous cells are grouped, including any gaps, and only clusters with more than 50,000 inhabitants are maintained as an "urban center";
- phase 3: all municipalities with at least half of the population within the urban center become candidates to be part of the city;
- phase 4: the city is defined ensuring that:
- there is political correspondence;
- at least 50% of the population lives in the urban center;
- at least 75% of the urban center population lives in the city.

Commuting Zone Commuting areas of the FUAs (Eurostat 2015) which are defined according to the following steps:

- if 15% of people with a job, who live in one city, work in another city, they are treated as a city;
- all the municipalities in which at least 15% of their residents with an occupation work in a city are selected;
- municipalities surrounded by a single functional area are included, discarding non-contiguous ones.

FUA (Functional Urban Area) Functional urban area, with a resident population greater than 100,000 inhabitants, consisting of its city—or city—and its commuting zone—or commuting zone. The identification of the FUAs represents an attempt by Eurostat (2015) to harmonize the definition of "metropolitan area", adopting the concept of the functional urban area, in which a significant proportion of residents move toward the city.

Indices and Statistics

ESAI (Environmental Sensitive Area Index) Synthetic index of vulnerability to land degradation, widely used in the Mediterranean area, which considers together numerous variables and thematic indicators concerning the climate, soil quality, plant cover and soil management, considered as significant factors in triggering land degradation. The ESA system is based on elementary papers (layers or layers) reworked through the attribution of scores capable of quantitatively expressing the weight exercised by each environmental parameter in modifying the equilibrium condition of an ecological system. The highest values, for example, refer to climatic, edaphic, vegetational or managerial conditions that exert negative pressures on the territory and predispose it to be vulnerable. About the calculation algorithm, in summary, the quality of the soil, climate, vegetation and socio-economic factors is estimated as the geometric mean of the scores of each layer.

Shannon Diversity Index The Shannon–Wiener H' index is a diversity index used in statistics in the case of groups with an infinite number of elements.

$$H' = -\sum_{j=1}^{s} p_j \log_e p_j$$

where pj is the proportion of the j-th element (Σjpj = 1) and s is the number of elements. Dividing H' by the maximum possible value H' max = loge (s), an index between 0 and 1 is obtained.

The expression of the diversity index is derived from the information theory developed by Claude Shannon, in particular from the definition given in it for the entropy of a discrete source.

Pielou Equitability Index An index that expresses the degree of homogeneity with which the elements are distributed in the various groups that make up a whole. Pielou's fairness index J takes into consideration the method of distribution of the individual elements in the various groups:

$$\mathrm{J} = \frac{H'}{\log_2 S}$$

where H' is the value of the Shannon–Wiener Diversity Index and S is the number of elements present in the given set. Equitability tends to 1 the more the elements are evenly distributed among the groups. It tends to 0 as much as some elements dominate numerically over others.

NDVI (Normalized Difference Vegetation Index) An indicator that is used to analyze the measurements obtained by remote sensing, typically but not necessarily from a special satellite, and to evaluate whether the observed area contains live vegetation.

NDVI is calculated as follows:

$$\mathrm{NDVI} = \frac{(NIR - VIS)}{(NIR + VIS)}$$

where VIS and NIR stand, respectively, for the spectral reflectance measurements acquired in the visible (red) and near-infrared regions.

Negative values of NDVI (approaching −1) correspond to water. Values close to zero (from −0.1 to 0.1) generally correspond to arid areas with rock, sand or snow. Finally, positive and low values represent shrubs and pastures (approximately between 0.2 and 0.4), while high values indicate temperate and tropical rain forests (values approaching 1).

SDI (Soil Degradation Index) The index developed as part of the DISMED project and validated in the field by Lavado Contador et al. in 2009, it allows to jointly evaluate the level of soil quality and that of its sensitivity to land degradation. Since 2003, the European Environment Agency has made available a raster of this index, which derives from the geometric average of the information layers relating to the soil, climate and vegetation quality indices:

$$\mathrm{SDI} = (\mathrm{SQI} \times \mathrm{CQI} \times \mathrm{VQI})^{\frac{1}{3}}$$

It can range from 1 (a value that corresponds to the maximum level of land quality and the minimum level of land degradation) to 2 (which, conversely, corresponds to the lowest level of land quality and the highest level of land degradation).

References

Adger WN, Hughes TP, Folke C, Carpenter SR, Rockström J (2005) Social-ecological resilience to coastal disasters. Science 309:1036–1039

Bruegmann R (2006) Sprawl: A compact history. University of Chicago press.

Couch C, Leontidou L, Arnstberg KO (2007) Introduction: Definitions, theories and methods of comparative analysis. In: Urban Sprawl in Europe: Landscapes, Land-Use Change & Policy, pp. 1–38

EAE (2006) Urban Sprawl in Europe. The Ignored Challenge, København

Eurostat (2015) European cities – the EU-OECD functional urban area definition. https://ec.europa.eu/eurostat/statisticsexplained/index.php/Archive:European_cities_%E2%80%93_the_EU-OECD_functional_urban_area_definition

Folke C, Carpenter SR, Walker B, Scheffer M, Chapin T, Rockström J (2010) Resilience thinking: Integrating resilience, adaptability and transformability. Ecology and Society 15(4):20

Holling CS (1973) Resilience and stability of ecological systems. Annual Review of Ecology Systems 4:1–23

ISPRA SNPA (2018) Consumo di suolo, dinamiche territoriali e servizi ecosistemici. Edizione

McCarthy JJ, Canziani OF, Leary NA, Dokken DJ, White KS (Eds.) (2001) Climate Change 2001: Impacts, Adaptation and Vulnerability. Cambridge University Press, Cambridge

UNCCD (1994) United Nations Convention to Combat Desertification. Parigi, UNEP

UNCCD (2015) Report of the Conference of the Parties on its twelfth session. Addendum (ICCD/COP(12)/20/Add.1)

United Nations Department of Economic and Social Affairs (2016) Sustainable Development Goals, Sustainable Development Knowledge Platform. https://sustainabledevelopment.un.org/?menu=1300

I. Tombolini et al., *Land Quality and Sustainable Urban Forms*, Springer Geography,
https://doi.org/10.1007/978-3-030-94732-3